T0091975

The Sounds of the Cosmos

The Sounds of the Cosmos

Gravitational Waves and the Birth
of Multi-Messenger Astronomy

Mario Díaz, Gabriela González, and Jorge Pullin

The MIT Press

Cambridge, Massachusetts | London, England

The MIT Press would like to thank the anonymous peer reviewers who provided comments on drafts of this book. The generous work of academic experts is essential for establishing the authority and quality of our publications. We acknowledge with gratitude the contributions of these otherwise uncredited readers.

This book was set in Stone Serif and Stone Sans by Westchester Publishing Services. Printed and bound in the United States of America.

Library of Congress Cataloging-in-Publication Data

Names: Díaz, Mario (Mario Claudio), author. | González, Gabriela, 1965– author. | Pullin, Jorge, author.
Title: The sounds of the cosmos : gravitational waves and the birth of multi-messenger astronomy / Mario Diaz, Gabriela Gonzalez, and Jorge Pullin.
Description: Cambridge, Massachusetts : The MIT Press, [2023] | Includes index.
Identifiers: LCCN 2022010112 (print) | LCCN 2022010113 (ebook) | ISBN 9780262544948 (hardcover) | ISBN 9780262372787 (epub) | ISBN 9780262372794 (pdf)
Subjects: LCSH: Gravitational waves. | Gravitational waves—Detection.
Classification: LCC QC179 .D53 2023 (print) | LCC QC179 (ebook) | DDC 539.7/54—dc23/eng20221025
LC record available at https://lccn.loc.gov/2022010112
LC ebook record available at https://lccn.loc.gov/2022010113

10   9   8   7   6   5   4   3   2   1

Contents

Preface   ix

1   Gravity from the Ancient Greeks to Newton   1

1.1 Physics in Ancient Greece   4
1.2 The Renaissance: Physics on the Earth   5
1.3 The Renaissance: Physics in the Sky   7
1.4 Waves and Action at a Distance   10

2   Einstein's Special Theory of Relativity   15

2.1 Electromagnetic Waves   15
2.2 The Speed of Light   17
2.3 A New Relativity   20
2.4 Implications of the New Theory   23
2.5 The Step   24

3   The General Theory of Relativity   29

3.1 The Next Steps   29
3.2 Physics Calls Geometry for Help   34
3.3 The Early Successes of the New Theory   44
3.4 The Shocking Predictions of the New Theory   47

4   Gravitational Waves: Their Long and Difficult History   55

4.1 The Prophecy   56
4.2 Waves in Physics   58
4.3 Gravitational Waves   59
4.4 The Sources of Gravitational Waves   61
4.5 Einstein's Doubts about Gravitational Waves   65

4.6 The Low Tide of General Relativity   68
4.7 Gravitational Waves Do Exist!   72

5   **The Life and Death of Stars**   77
5.1 The Birth and Life of Stars   77
5.2 Small Masses: White Dwarfs   79
5.3 Medium Masses: Neutron Stars   82
5.4 Extreme Masses: Black Holes   84

6   **Astrophysical Sources of Gravitational Radiation**   89
6.1 Stellar Binary Systems   89
6.2 Rotating Stars   92
6.3 Supernova Explosions   93
6.4 Stochastic Background of Gravitational Waves   94

7   **Numerical Relativity**   97
7.1 First Attempts   98
7.2 The Grand Challenge   99
7.3 The 2005 Surprise   100
7.4 Expectations and Reality   101

8   **A Brief History of Terrestrial Gravitational Wave Detectors**   105
8.1 The Small Amplitude of Gravitational Waves   105
8.2 Bar Detectors   109
8.3 Interferometric Detectors   112

9   **The Technology of LIGO**   123
9.1 Interferometry: Differences of Length   123
9.2 No Air Allowed: Vacuum System   127
9.3 Suspended Mirrors   130
9.4 Seismic Isolation   133
9.5 A Quantum Enemy: Shot Noise   136
9.6 A Quantum Friend: Squeezing the Vacuum   138
9.7 Optical Tricks and the Need for Control   140
9.8 Thermal Noise: If It's Warm, It Moves   144
9.9 Summing Up   147

10   **At Last: Detections—and Many!**   151
10.1 The First Detection: GW150914   153
10.2 The Second Detection: GW151226   161
10.3 More Detections: Catalogs of Gravitational Waves   163
10.4 Even More Detections: New Catalogs   165
10.5 The Properties of the Detected Sources   166

11 **The Birth of Gravitational Wave Multi-Messenger Astronomy** 171

11.1 The Network of Detectors   171

11.2 Multiple-Messenger Astronomy   173

11.3 The Mystery of the Gamma Rays   176

11.4 All That Glitters . . .   179

11.5 The Speed of Expansion of the Universe   181

12 **The Future** 185

12.1 Growing the Terrestrial Detector Network   185

12.2 Even More Advanced Detectors   187

12.3 Even Longer Detectors: In Space   191

12.4 Detecting Primordial Gravitational Waves   194

Epilogue 197

Acknowledgments 199

Notes 201

Recommended Reading 205

Index 207

Preface

On February 11, 2016, at the National Press Club in Washington, DC, the LIGO Scientific Collaboration announced the discovery of gravitational waves. A simultaneous, similar announcement was made by the Virgo Scientific Collaboration in Europe. In different roles, the authors of this book were all present at the announcement in Washington.

Those waves were predicted by Albert Einstein in 1916 and are a new way to study the universe: we can see the cosmos with light, and we can "hear the sounds" of the cosmos with gravitational waves.

The purpose of this book is to bring the subject of gravitational waves to the general public and to tell the story of how it was possible to discover and understand these waves. We describe the physics developments that led to this achievement and reflect the eternal quest of humankind to understand the universe.

1

Gravity from the Ancient Greeks to Newton

Some of the earliest recorded attempts to understand our surroundings in a quantitative manner were undertaken by the ancient Greeks. However, our modern understanding of physics begins with Isaac Newton. This great span of time between the efforts of the Greeks and those of Newton gives a sense of how nontrivial this undertaking has been.

In 1665 Isaac Newton was finishing his studies at the University of Cambridge, and at the same time the bubonic plague was breaking out in England. The Great Plague (1665–1666) was the worst outbreak of plague in England since the black death of 1348.

The pandemic forced the university to close, and between the summer of 1665 and the spring of 1667, Isaac Newton went back to his home in Woolsthorpe by Colsterworth (in the East Midlands of England). The solitude gained by the forced quarantine allowed the young Newton to concentrate on his plans to study "matter, place, time, and motion . . . the cosmic order, then . . . light, colors, vision" (according to his biographer Richard Westfall).[1] Another quarantine, brought about by a modern plague (the Covid pandemic), gave us some time to accomplish a more modest endeavor: to expedite the writing of this book.

Legend has it that while Newton was sitting under an apple tree in his backyard, a falling fruit knocked him on his head, and this bonking experience was Newton's eureka moment (see figure 1.1). He realized that the force that attracted the apple toward the Earth was the same one making the Moon move around the Earth. Although there is no historical evidence that this apple event actually took place (the story was told by others many years after the fact), the apple tree that might have inspired Newton exists and survives in Woolsthorpe. Sprouts from the tree grow at various physics institutes around the world.

Newton's theory of gravity is now taught as part of the standard science high school curriculum. It establishes that two bodies attract each other with a force that is proportional to their masses; that is, the more massive or "heavy" each of these objects is, the stronger the force of attraction. What also determines the force is the distance between the objects, but in this case the relationship is opposite to the role played by mass: the attraction diminishes if the distance increases. The attraction goes as the inverse of the square of the distance: if two bodies move three times as far apart, the attractive force will be nine times smaller.

Despite its simplicity, Newton's theory of gravity is a towering intellectual achievement. Matter exerts an attraction on any other matter around it. This attraction follows a clear and precise mathematical prescription, what physicists call a "physical law." This law is also universal: it describes everyday terrestrial phenomena of our immediate surroundings, like the fall of an apple, as well as celestial events, like the motion of the planets and the Moon. Newton's law makes no distinction between the heavens and the Earth: there is only one physics governing events in both realms. One of the characteristics that distinguishes contemporary physics is its attempt to unify the various theories that explain different aspects of the universe. This equalizing of the mundane and the celestial, achieved scientifically

Figure 1.1

One of the first proposed logos for the Apple Computer Company depicted Newton under the apple tree. By Ronald Wayne. (Credit: Wikimedia Commons, public domain.)

by Newton, is probably the first great theoretical and philosophical unification in the history of humankind. Let us, however, take a brief detour through the prehistory of physics: the ancient Greeks.

1.1   Physics in Ancient Greece

The first efforts to systematically develop theories explaining the nature of the physical world started with the Greek philosophers. Newton's theory of gravity in some sense recaptures the scientific and universal spirit of the classical Greek schools in the fifth and fourth centuries BC, before Aristotle, who believed that natural phenomena could only be explained by natural causes ("laws") and not by divine intervention.

Ancient Greek scientific thinking reached its pinnacle with Aristotle, who wrote about multiple aspects of human creativity, from the arts to the sciences. He was the first to write a book expounding ideas about the way nature works, which he titled *Physics* (in Greek "phusis," meaning "nature"). However, departing from the thinking of the older Greek philosophers, for him the physics of the Earth was different from the physics of the sky. Earthly objects were formed by four basic elements in different proportions: air, fire, water, and earth. A rock, for instance, contains much more earth than it does water, air, or fire. Thales had already proposed that water was a fundamental element that was part of all substances.

For Aristotle, celestial bodies were of a different nature than everyday objects: that is, they were made of different elements than the objects on the Earth. Celestial bodies were composed of a different kind of matter, one which he named "quintessence." Aristotle believed that the heavens—understood as the sphere where the Sun, Moon, and planets move—were immutable and of a different essence than the Earth.

One of the fundamental goals of his physics was to explain the movement of the bodies on the Earth: Why does motion occur? Why does an object appear to fall faster the heavier it is? If we throw an object, can we predict where it will fall?

For Aristotle, movement existed because every object has a "natural place," moves toward it, and when there, remains still. If a rock is lifted and then released, it will inevitably fall to earth. According to Aristotle, who did not have a theory of gravity to explain it, the reason for this phenomenon was that the stone has a much larger component of earth than water, fire, or air. Therefore, when it is taken from its natural place—the ground, or equivalently the Earth—the rock "wanted" to get back to the ground and stay there.

For Aristotle's physics, rest was the fundamental state of physical bodies. (Although intuitively this sounds true, we now know that it is false!) This static and hierarchical vision of nature was influential in Western thought until the Renaissance, not only in scientific circles but also in philosophical and religious ones. The Aristotelian version of physics dominated Western thinking for more than 1,500 years.

1.2   The Renaissance: Physics on the Earth

In the second half of the sixteenth century, two men were born within 7 years of each other who brought about an upheaval in the foundations of Aristotelian physics: Galileo Galilei and Johannes Kepler. Both made important contributions based on experiments and observations, but with different motivations.

Galileo wished to understand the fall of objects toward the ground: If an object falls from a given height, how long does it take to fall? With great ingenuity and a great ability for abstraction never before applied in science, Galileo observed that the speed of objects in free fall increased with time: they accelerated. Experimenting with balls of the same size but different materials and weights, he also discovered that this acceleration was constant and independent of the mass of the objects. This observation predicts a cannon ball and a feather fall at the same rate: Can this be true? In everyday experience, this does not happen due to air resistance, but it does happen in a vacuum. In a popular NASA video, David Scott—commander of the Apollo 15 mission—drops a hammer and a falcon feather on the surface of the Moon (where there is no air resistance), and both objects hit the surface of the Moon at the same time. Galileo also noticed that rest was not a "natural state": In a vacuum, a body moving at constant speed keeps moving in such a way if no force acts on it. This is known as the *law of inertia*. Galileo ended forever the Aristotelian conception of rest as the natural state of bodies; motion at a constant speed was as natural as being at rest. Objects remain in their initial state unless forces act on them.

An important consequence of Galilean physics is that if one observes an object moving at a constant speed from a vehicle moving at the same speed and in the same direction, the object will be perceived as if it were at rest. The concept of "motion" (at least at a constant speed) is not absolute, because there will always exist a *reference frame* (a frame from which to measure distances in space) in which the object can be described as being at rest. If this idea appears rather abstract and confusing, just think of the feeling we get in a stopped train when we see another train pass by in the track next to ours: For a fraction of a second, we think that our train is the one that is moving. Or when at a traffic stop, if the vehicle stopped next to ours lunges forward before ours, for a second it looks as if our car goes backward.

With Galileo, modern mechanics was born. But he also pioneered a new way to do and think about physics: the modern "scientific method." This method was brilliantly used by Newton first and Einstein later.

1.3   The Renaissance: Physics in the Sky

Kepler, 7 years younger than Galileo, was trying to understand the mystery of the motion of planets in the sky. The planets, a term that in Greek means "wanderers," were given that name by the ancient Greeks because these objects were the only "stars" in the sky that had a significant motion, changing position not only throughout the night but during the year. That is, if one imagined the sky as rotating around the Earth, the planets' displacements were different from those of the rest of the stars.

Until Copernicus—born at the end of the fifteenth century, during the rise of Renaissance—the geocentric model was predominant. In this model, the planets move in circles around the Earth. But it was very difficult to explain with this model the movement of the planets: during a certain period of the year, they seem to move "backward" until they eventually start moving forward again, creating a strange imaginary curl in the sky (see figure 1.2). This apparently backward trajectory that planets seem to have for a few days is known as retrograde motion and is more visible for planets that are farther away from the Sun than the Earth is.

Long before Copernicus, in the second century AD, during the Roman Empire, the mathematician and astronomer Ptolemy perfected a model proposed centuries before to explain the retrograde motion of the planets, in which the Earth was the center of the solar system. In this theory, Ptolemy stated that the planets move in circles around imaginary points in the sky, and these points, in turn, described other circles around the Earth called "epicycles." This model—although it is complicated—was sufficiently capable of making predictions and fitting observations of that time. Ptolemy's great work—the *Almagest*—was the most respected astronomy book for almost thirteen centuries. Although few women are known to have been involved in math or astronomy in antiquity, Hypatia from Egypt was an exception: she edited the existing text

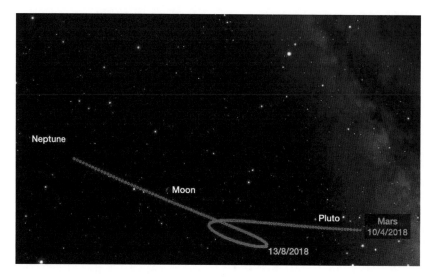

Figure 1.2

Mars' retrograde motion as seen in the sky over Buenos Aires from April 10 until December 15, 2018. Each dot along the path represents the position of the planet in the sky at the same time of night. Note that on June 25 (at the apparent intersection of the curve with itself), the planet starts to move apparently backward until August 13 of the same year, when it resumes its forward motion. (Simulation made by the authors with the computer program TheSky, Software Bisque.).

of Book III of Ptolemy's *Almagest,* explaining the motion of the Sun and the epicycles. Subsequent observations, however, showed that Ptolemy's model was wrong.

When Kepler studied math in Germany, the Sun-centric ("heliocentric") model of Copernicus had acquired widespread acceptance. In this model, the planets go around the Sun. If the planets that are closer to the Sun than the Earth is move faster, then it was easy to understand that the ones farther away—which move slower— appear to move backward when seen from the Earth. But why do the planets closer to the Sun move faster? What is the shape of their trajectory around the Sun?

Moved by a religious mystique, Kepler thought that God had implemented a rational plan for the universe; that is why he described his quest as "reading God's mind." He also thought that a good place to start that reading was to explain the motion of the planets. Kepler had been remarkably influenced by the mathematical ideas of Pythagoras and the classical Greek schools, which saw in mathematics—and particularly geometry—the most profound expression of an intelligible universe. For many years, Kepler thought that the planetary orbits were circles and that they should obey a simple geometric relation between them.

Contemporaneous with Kepler, Tycho Brahe, a Danish astronomer, managed to observe with great precision Mars' orbit. Kepler, having unsuccessfully worked with the models based on his mystical beliefs, decided to limit himself to describe in the most objectively possible way the trajectories of the planets. In particular, he wanted to find relations between the velocity of the planets and the size of their orbits. After several years of attempts to model the orbits with circles, he tried other geometric shapes and eventually established what today we know as Kepler's first law: the planets move on ellipses. Kepler also found that the planets moved faster the closer they were to the Sun: this is Kepler's second law. Finally, he arrived at his third law, which involves a precise relation between the time taken by a planet to complete its orbit around the Sun and its distance from the Sun.

If we consider Galileo to be the founder of modern physics, the development of the theory of planetary motion by Kepler became the paradigm of modern astronomy. His eagerness to adjust his theory to the observations of Tycho Brahe instead of to his religious beliefs embodies well one of the thoughts of Carl Sagan, the famous US scientist who popularized astronomy through the TV show *Cosmos*. Sagan said that a scientist must keep an open mind, independent of prejudice and with a dose of skepticism, and that ideological or religious preconceptions—although they surely condition human thought—must be left behind in favor of the critical examination of nature. Any theory can only be accepted if it can explain observations and make verifiable predictions. An honest scientist should be ready to question even the theory she dearly cherishes if it is refuted by experimental evidence.

In general, when theories developed in the modern era of physics showed limitations—for instance, in their predictive capacity—it was not necessary to abandon them completely. The new theories usually represent an improvement on old ones, but they keep elements of their predecessors. One of the remarkable examples that we will discuss in chapter 3 is Einstein's theory of gravity, which makes predictions similar to Newton's theory when bodies are not moving too fast and fields are weak.

With his theory of gravity, Newton was able to unify the terrestrial mechanics of Galileo and the celestial mechanics of Kepler. He not only explained gravity, but he also elaborated a theory of motion. The success of this theory has been extraordinary. It describes in a simple and precise way the motions of interacting bodies. If their positions and velocities at a given time are known, then it is possible to predict their positions and velocities at any future instant.

The success of Newtonian celestial mechanics was spectacular: it predicted the motions of all planets in the solar system to very good precision. But the orbit of Mercury, the closest planet to the Sun, could not be completely explained by this theory. Mercury's orbit has a small discrepancy with Newton's prediction. The planets' perihelion (the point of the orbit closest to the Sun) moves, and Newton's prediction for the movement of Mercury's perihelion was slightly different (a bit less than 1 percent) than the observed value. Modern physics and the mechanics of Kepler, Galileo, and Newton left the nineteenth century a small mystery. Resolving it would require a new theory: Einstein's general relativity.

1.4 Waves and Action at a Distance

In the nineteenth century, the Scottish physicist James Clerk Maxwell unified the theories that explain electricity and magnetism, creating the theory known as electromagnetism. It predicts "electromagnetic waves" propagating at the speed of light, approximately 200,000 miles per second.

This theoretical discovery motivated experimental physicists like Heinrich Hertz to produce and observe the electromagnetic waves in the lab. Up until then, waves had been a phenomenon associated with acoustics and fluids.

The force of gravity as formulated by Newton was in this sense qualitatively different: it was an instantaneous force with an infinite speed of propagation. This seemed unnatural to Einstein and many others before him. Shortly after the formulation of the general theory of relativity, Einstein discovered that his equations had certain types of solutions that represented waves propagating in space. They traveled at the speed of light, like electromagnetic waves, and his theory predicted that no interaction propagates faster than that speed. The gravitational waves can be considered the messengers of gravity.

A finite speed for gravity had been considered 100 years before Einstein, through a series of historical coincidences. It started with British theologians, who were upset with Newton's theory, because it described the planets in the solar system as following orbits indefinitely without divine intervention. This model looked very similar to the atheist ideas of French philosopher René Descartes. He claimed that it was not necessary to appeal to theological or divine notions to explain nature. His philosophy, which develops the foundations of modern rationalism, did not need a God to explain natural phenomena. Nothing possibly upset theologians more than ideas that get rid of a supreme being at the center of the functioning universe. Newton tried to reconcile with the theologians, attempting to include the need of divine intervention through the study of the motions of the Moon, which were not well understood. The distance between the Earth and the Moon appeared to change.

In addition to these strange theological motivations, to understand with precision the motion of the Moon would have great practical importance. The sixteenth century was the beginning of transoceanic navigation involving routes far away from the coast. For such navigation, it is important to establish with accuracy the latitude and longitude of a ship. Latitude measures the position of an object either north or south of the equator. Longitude measures the position of the object to the east or the west of a meridian that is chosen as a reference (figure 1.3), just as the equator is chosen for latitude. To determine the latitude, one only needs to see how far up in the sky the Sun rises during the day. To determine longitude,

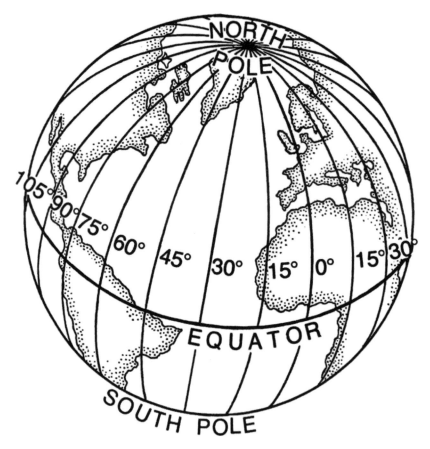

Figure 1.3

Longitude lines on the globe. (Credit: Pearson Scott Foreman, Wikimedia Commons, public domain.)

one needs to know the time at which the Sun is highest in the sky. The clocks that could be used to determine longitude with enough accuracy at that time used pendulums (like a grandfather's clock). But aboard a ship sailing on the ocean, the rocking motion makes pendulum-based clocks very inaccurate. Another way to determine the time is to use a telescope to follow stars. But that is also very difficult to do in a moving boat. Given its larger apparent size in the sky, observing the Moon is much more feasible. If its motion could be predicted with accuracy, one could determine what time it is by determining the position of the Moon in the sky.

The British Parliament and the French National Academy of Sciences had created prizes for anyone who could solve the problem of establishing the longitude of ships. The British Parliament even created the Board of Longitude to administer the prizes, which taken together were on the order of several million dollars in today's currency. The board called them "prizes," but in reality, they were the first system of government-based scientific funding, as the prizes were awarded for partial advances. At the time, there were no other sources of government support for researchers, so the prizes attracted some of the most brilliant minds of the time to the problem. Among them was the French mathematician Pierre-Simon Laplace.

Laplace wondered what would happen if gravity propagated at a finite speed. One object orbiting around another would feel the other's gravity with a delay with respect to its current position, since gravity would take a finite time to reach from one object to the other. That would alter the orbits that Newton's theory predicted. In the end, Laplace's explanation for the motion of the Moon turned out to be incorrect. The effect is due to tides, in which both the Earth and Sun intervene. But Laplace was the first person to consider that gravity could propagate at a finite speed. In fact, the calculation Laplace made is what in modern parlance researchers call "reaction to gravitational radiation," and it is an active field of research today in a different context.

The alteration in the orbits that Laplace noted are because the systems emit gravitational waves, which take away energy from the system. To understand these concepts properly, one needs a better theory of gravity. That theory is Einstein's general theory of relativity, which was proposed by him in 1915, 110 years after Laplace's early calculation.

Before introducing his general theory, Einstein developed the special theory of relativity, which we will discuss in chapter 2.

2
Einstein's Special Theory of Relativity

The towering intellectual achievements that represented Newton's theory of gravity and Maxwell's theory of electromagnetism, and their great success in explaining physical phenomena, gave the impression at the beginning of the twentieth century that research in physics was "complete." That is, apparently there was no fundamental principle left to be understood or explained. But there were still questions, several of which attracted the attention of Einstein. One of them was that when speaking of the velocity of an object, it is always measured relative to another one. But in Maxwell's theory, light always moves at the same speed with respect to all observers, regardless of their relative state of motion. The light emitted by a stationary lamp or the headlight of a moving train both move at the same speed. This property of electromagnetism seemed incompatible with the physics of motion created by Galileo and Newton. Einstein proposed in 1905 a theory that solved this problem, creating a new way of understanding "relative" motion: the theory of special relativity.

2.1 Electromagnetic Waves

Maxwell's formulation of electromagnetism in 1865 was a second "unified" theory of physics after Newton's theory of gravity (explaining motion on Earth and in the sky). The theory, putting together electricity and magnetism, predicted the existence of electromagnetic waves, moving at the speed of light: 299,792 km/sec. Did this mean that light itself was a wave of electricity and magnetism?

Christiaan Huygens, a Dutch mathematician, physicist, and astronomer, thought light was a wave. He explained the observed effects of reflection and refraction and published his theory in 1690. Newton thought light was made of colored particles (explaining why white light was decomposed into colors by a prism), writing a book on optics in 1704 that included his "particle theory of light." Many debates followed, which ended in 1807, when Tomas Young demonstrated that light channeled through two pinholes made on a blocking shade produces interference patterns just like those between the waves produced by two stones thrown into a pond as they interact with each other. The phenomenon of interference is an essential component of gravitational wave detectors that we will discuss in chapter 8 and later chapters.

The electromagnetic waves predicted by Maxwell's theory are the waves that transmit radio and television—of course unknown at that time—but also light, which is "made" of electromagnetic waves, as are x-rays and gamma rays. The different names for electromagnetic waves are based on the wavelengths they can have (or equivalently, on their characteristic frequencies). They can be compared to waves in the ocean: the height of the wave is called the amplitude and the distance between two crests is the wavelength. The more frequent the arrival of the crests, the shorter the length between successive crests: the frequency of the wave is inversely proportional to its wavelength.

The transmission and reception of radio waves was demonstrated by Heinrich Hertz in 1887 in Germany. A few years later, in 1901, Guglielmo Marconi spectacularly transmitted them across the Atlantic from Ireland to Newfoundland in Canada. It is interesting to note that when his students asked Hertz about the usefulness of his experiment, he replied: "It's of no use whatsoever. This is just an experiment that proves Maestro Maxwell was right—we just have these mysterious electromagnetic waves that we cannot see with the naked eye. But they are there."[1]

2.2   The Speed of Light

Maxwell's theory predicts the speed of light in terms of constants that were known from experiments with electricity and magnetism, and it was a surprise that the predicted value agreed with observations.

Measuring the speed of light has been attempted ever since humans started pondering about the laws that govern the universe. In ancient times, it was thought that the speed of light was infinite. Galileo was one of the first physicists to attempt to measure it. He proposed the following experiment: two persons located at a predetermined distance from each other would each carry a lamp, covering it and then briefly uncovering it. Measuring the time that the light took to get from one lamp to the other and back, its speed could be calculated. But Galileo concluded that the experiment only yielded a highly inaccurate result, because light travels very fast. He estimated that it travels ten times faster than sound (it actually travels almost a million times faster).

In 1676 the Danish physicist Ole Rømer was able to make the first reliable measurement of the speed of light. He noted that the eclipses of the four largest moons of Jupiter—which can be observed with a good pair of binoculars—happened earlier than expected when the planet was closer to Earth, since light had to travel a shorter distance. Knowing the distances between the planets, one can estimate the speed of light with some accuracy.

The speed of light was also measured using *stellar aberration*. Due to the motion of the Earth around the Sun, the angle at which stars appear in the sky changes throughout the year: this is known as parallax (as shown in figure 2.1). Parallax also explains what happens when we place a finger in front of the nose and observe it with one eye closed and then with the other eye closed. The finger will appear to move with respect to the background (we are seeing the finger from two different perspectives).

James Bradley, an English Royal Astronomer of the seventeenth century, attempted to measure the parallax of the star known as Gamma Draconis. He noted a substantial difference from the expected result: the angle changed, but the difference did not correspond to the position of the Earth in its orbit—it was delayed. This is why this effect is known as stellar "aberration." What is its origin? Light takes some time to arrive at Earth, but during its transit time from the star, the Earth has moved. Bradley was able to estimate the angular change that this effect implies, which depends only on the speed of light and the speed of the Earth.

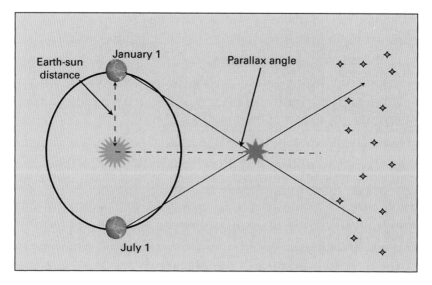

Figure 2.1
Parallax to a nearby star.

Armand Hyppolyte Louis Fizeau, one of the great French physicists of the nineteenth century—his name appears with those of 71 other scientists and engineers at the base of the Eiffel Tower—obtained the best measurement of the speed of light at the time, with an accuracy of about 5 percent. He measured the speed of light in 1849 in an experiment using a beam of light traveling 8 km, being reflected by a mirror and using a toothed wheel rotating at just the right speed to let the initial beam of light go through the wheel and to block the returning beam. Fizeau also measured the speed of light traveling in media like water: it was slower. This result was expected, because it was known that the speed of light traversing a physical medium (like water or a glass) is slower than in vacuum. But Fizeau tried also to measure the speed of light in water when water was in motion. And he measured it in both directions: in the direction of the water flow and also in the opposite one. He found a slight variation in both speeds but not the one expected from using the addition of velocities of Galileo. How could this be? This mysterious result was nevertheless consistent with Bradley's observations. In both cases the speed of light did not seem to be affected by the relative motion of the observing systems.

As we discussed, Maxwell's theory predicted electromagnetic waves. But scientists wondered what these waves traveled "on." Water waves and sound waves are the motion of water and air particles: sound does not travel in vacuum. For the physicists of that time, light similarly should require a medium to propagate. They theorized about a mysterious "luminiferous aether," which would have been the propagating medium. The belief was that the universe was full of this special substance, and scientists wondered whether it was quiet (at rest) or not. Was it dragged by the Earth as it rotates?

Perhaps measuring the speed of light in directions parallel or opposite to that of the rotation of the Earth could say something about the aether's motion. Between April and July 1887, the American physicists Albert Michelson and Edward Morley carried out an experiment that attempted to measure the relative motion of light with respect to the aether. The Earth orbits around the Sun with a speed of 30 km/s (108,000 km/h). Scientists at the time advocating for the aether theory predicted two possible outcomes from the experiment: (1) the aether was stationary and affected only in a very limited way by the motion of the Earth, or (2) the aether was completely dragged by the Earth. If the first hypothesis were true, an "aether wind" would affect the experiment, if the second were true, it would not. The experiment determined that the speed of light, when coming from a source emitting in the direction of the rotation of the Earth, was the same as when the source emits it in the opposite direction. Thus, either the aether was being dragged by the Earth, or the velocities of light and aether could not be added in the usual way.

2.3   A New Relativity

Galileo had postulated that the motion of a body having constant speed was indistinguishable from one at rest, since it is always possible to find a reference frame in which the moving object appears at rest. This is the first formulation of a theory of relativity, called Galilean relativity.

A simple example provides a clear visualization of this concept. María and José are walking together at the same pace at an airport and reach a moving walkway; María steps on it, while José keeps walking on the floor next to the walkway. Although they keep walking at the same pace, we see María moving ahead of José: the walkway speed is added to hers.

Galilean relativity states that the speed of an object with respect to a system that also moves—like the walkway—is calculated by adding the speed of the object plus the speed of the system. This principle of addition of velocities, which forms part of classical mechanics, had been solidly demonstrated and verified.

But as we mentioned, there were some observations contradicting this way of adding velocities. Einstein himself said that he was motivated to develop his relativity theory by two experiments: Fizeau's measurement of the speed of light in a medium in motion and the stellar aberration phenomenon of Bradley.

Einstein published the special theory of relativity, in contrast to Galilean relativity, in 1905, in a paper called "On the electrodynamics of moving bodies." That year was his "annus mirabilis," in which Einstein published four papers that changed the history of physics in more than one way.

In his paper, he defined "inertial reference frames," which move at constant speeds between them, as being the appropriate systems from which to measure distances and times. That is, they do not experience an acceleration or change in their velocity. He then proposed two fundamental postulates about inertial reference frames: (1) the laws of physics are the same in any inertial reference frame; and (2) the speed of light has the same value when measured in all inertial reference frames, independent of their relative velocities.

What is the meaning of these postulates? The first one states that physical phenomena that are described in a reference system do not change when one describes them in another system that moves at a constant speed with respect to the first. For instance, the force of gravity perceived by an observer who drops a pair of binoculars from atop the mast of a fast ship is the same as that perceived by an observer on a pier who sees the ship pass by her. The trajectory observed from the ship and the pier differ, due to the displacement of the ship with respect to the pier. For the observer on the ship, the binoculars fall vertically, whereas for the observer on the pier, it will describe a curved trajectory (in both cases the binoculars end up at the bottom of the mast). The physical law—the attraction of gravity—does not change. More to the point of the title of Einstein's paper, Maxwell's laws of electromagnetism are the same in all inertial frames.

The second postulate says that Galilean relativity (the addition of velocities) cannot be applied when light is involved. This second postulate has crucial implications. Moreover, in this context, the aether—an absolute reference frame—is not needed.

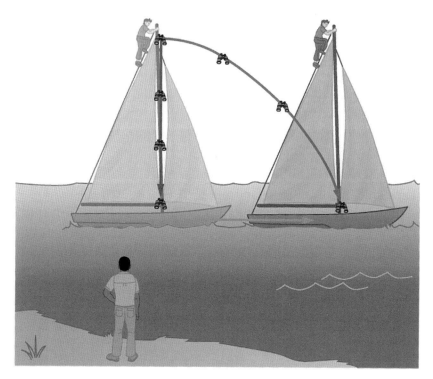

Figure 2.2

A pair of binoculars dropped in a moving ship, as seen from the ship and from the pier. (Credit: College Physics, Paul Peter Urone, Roger Hinrichs. Published by Open-Stax. License: CC BY: Attribution.)

Contrary to the most common historical presentations of these developments, Einstein never referred in his special relativity formulation to the results of Michelson and Morley. In today's typical physics courses, a bit of history is discussed, and it is commonly stated that the results of Michelson and Morley led Einstein to formulate his theory. But Abraham Pais—who wrote the most complete scientific biography of Einstein—says that there is no written record indicating that Einstein was influenced by the experiment. It is likely he knew about it, but he always quoted Fizeau's experiment and the measurement of aberration of starlight as his sources of inspiration.

Nevertheless, it is also clear that what Einstein had in his mind ruled out the existence of an aether; he was also keenly aware of the mathematical calculations that provided a logical framework for his special theory. Hendrik Lorentz (a Dutch physicist who won the Nobel Prize in 1902) and George Fitzgerald (an Irish physicist) showed that Fizeau's experiment and the constancy of the speed of light could be understood with a mathematical formulation of the addition of velocities different from the Galilean one. Einstein's genius consisted of promoting the result of the constancy of the speed of light to a universal law. At the same time, he clearly discarded the assumption of the existence of the aether as a useless historical artifact.

2.4 Implications of the New Theory

With special relativity, the absolute notions of times and distances of Newton's theory disappeared. Light has a speed—in vacuum—that is always the same, independently of the observer who measures it (and of the source that emits it). Since speed is distance divided by time, if the speed of light is constant, then necessarily times and distances must change such that different observers who are in motion with respect to each other measure the same speed of light.

The implication was that the time measured by an observer—that is, the time the observer reads on her watch—is different from the time that another observer reads on his own if he is moving with respect to the first observer. In addition, the distances—the size of an object—are different if they are measured in the direction of motion of two systems of reference with different speeds.

Another very surprising consequence is that simultaneity is a concept without an absolute sense: there is no clock that measures universal time. Two events that are simultaneous for an observer will not be so for another one that moves with respect to the first. This is in flagrant contradiction to our everyday intuition, but this intuition is based on speeds that are much lower than the speed of light.

To relate positions and velocities in both systems required new formulas. The new mathematical equations that relate lengths and times between both systems are known as *Lorentz transformations*. The formulas proposed by Lorentz and used by Einstein applied to reference frames in motion with respect to each other, where measurements of distances and times changed: these are now called "Lorentz contraction" and "time dilation," respectively. Given that the Lorentz transformations combine space and time, putting them on an equal footing, it is natural to think of a space-time of four dimensions, in which time appears as a "peer" of the three spatial dimensions (up/down, left/right, forward/backward). The structure of this space-time continuum got a name: Minkowski space-time. Hermann Minkowski, a German mathematician, published in 1908 an article similarly titled to (and based on) Einstein's article: "The fundamental equations for electromagnetic processes in moving bodies."

2.5   The Step

In spite of their simplicity, the two postulates of the special theory completely changed fundamental concepts of the physical world and affected other areas of knowledge, including philosophy.

Time—the natural measure of change—now becomes observer dependent and is no longer an absolute concept. Absolutes have had great importance for humans, since they provide certainties. The questioning of absolutes has a rattling effect on human belief systems. The impact of Einstein's theory was enormous and attracted the interest of intellectuals in very different fields of knowledge. On April 6, 1922, the French Society of Philosophy invited Einstein to speak about relativity. An important debate took place with the French philosopher Henri Bergson. At the time Bergson was more famous than Einstein, and there is speculation that Einstein did not get the Nobel Prize for the theory of relativity due to Bergson's opposition (Einstein received the Nobel prize in 1921 for his work on the photoelectric effect, one of his four papers of the annus mirabilis). In a certain sense, this debate implied a parting of the waters between physicists and philosophers in the subsequent years.

There is no doubt that the new theory of special relativity introduced a new way of thinking, at least in physics. It allowed applying the methods of classical mechanics to electromagnetism. And by doing this, it produced a new mechanics: relativistic mechanics. A result that Einstein himself anticipated in 1905 is the equivalence between mass and energy. The energy of a body at rest is equal to its mass multiplied by the speed of light squared (the super famous $E = mc^2$ formula). Since the speed of light is enormous, even a stationary body with a small mass has a considerable amount of energy: this is known as its rest energy. Einstein proposed this equivalence between mass and energy in another one of his articles published in 1905 titled "Does the inertia of a body depend on its energy content?"

Near the end of the article, Einstein mentions that the prediction could be tested with a body that loses energy by radioactive emission. Einstein proposed that by measuring the energy emitted by a radioactive compound, like uranium salts, and weighing its mass before and after the emission, the formula $E = mc^2$ could be verified experimentally.

Radioactive disintegration is the process by which an unstable atomic nucleus loses energy when it emits electromagnetic radiation, and physicists, well aware of the principle of conservation of energy in nature, encountered a true puzzle at the time: Where did the energy come from? Radioactivity was discovered by Antoine Henri Becquerel in 1896 and was investigated by Pierre and Marie Curie in the years that followed, all three receiving the Physics Nobel Prize in 1903. In 1899, Marie Curie wrote: "The radiation [may be] an emission of matter accompanied by a loss of weight of the radioactive substances." It took almost three decades for experiments to have enough precision and the use of Einstein's formula to explain energy in radioactive phenomena, and the experiment was described by *Time* magazine in 1933: "In Paris last year the Curie–Joliots bombarded a piece of lithium with alpha particles, produced neutrons and boron atoms. . . . The significance of the boron atoms in the Curie-Joliot experiment attracted less attention until last week Dr. Kenneth T. Bainbridge, who weighs atoms at Bartol Research Foundation laboratories in Swarthmore, PA, presented an interpretation." Bainbridge's article was titled "The equivalence of mass and energy." Irene Joliet-Curie (daughter of Pierre and Marie Curie) and her husband, Frédéric Joliot-Curie, received the Nobel Prize in Chemistry in 1935 for their discoveries about radioactivity.

According to Pais—the Einstein biographer—Einstein wrote in 1905 to his friend, the Swiss mathematician Conrad Habicht, about his formula: "This line of thought is . . . fascinating, but I cannot know if the dear Lord, playing a prank on me, laughs about me with this discovery." At that moment Einstein could not imagine all the consequences of this incredible discovery, involving atomic energy and nuclear bombs.

In 1939, more than 30 years after writing his mass-energy paper, Einstein sent President Franklin Delano Roosevelt a famous letter, in which he mentioned that the technology to start a nuclear chain reaction was in the hands of the Nazis in Germany. He argued that the US government should invest money and support the development of an atomic bomb in the US. On July 16, 1945, the first test of this bomb took place in the middle of a desert called "Jornada del Muerto" (Journey of the Dead, in Spanish) in a stretch of land between Las Cruces and Socorro, New Mexico (the name refers to the difficult trail through the region). The energy liberated in the explosion—equivalent to 21,000 tons of TNT—confirmed soundly the principle of the equivalence of mass and energy predicted by Einstein and started a new era in the history of humankind. Einstein would later regret sending that letter.

Pais calls the task of understanding what was going through the mind of the thinkers who were changing the way of viewing the universe "the edge of history." Pais worked with Einstein at the Institute for Advanced Study in Princeton, NJ, and had the chance to

Figure 2.3
Cover of *Time* magazine in July 1946.

explore closely the sharpness of this edge. He states that when he himself asked Einstein about the evolution of thinking by scientists like Lorentz, Poincaré, and others during this transition time, Einstein replied that the birth of special relativity was "den Schritt" ("the step," in German). For Einstein, the development of special relativity—a revolutionary event in the history of human thought—with multiple scientific and cultural impacts, was only that: a first step in the development of a more complete theory.

Einstein envisioned his development of the theory of relativity within a unified theory that could explain the universe in its entirety. A theory that integrated all the fundamental ideas of physics in a single and elegant formula: a "theory of everything," as it was called by Richard Feynman, the great American physicist who won the Nobel prize in 1965.

Einstein made several attempts from 1905 to 1907 to integrate gravity with the theory of relativity. The theory of general relativity as we know it today was presented in 1915. For Einstein, the 10 years from 1905 to 1915 were a period of transition, which was also reflected in his professional life. He passed the exam that allowed him to teach, and he accepted a job as a college professor, leaving his staff position at the patent office. Einstein had married a college classmate, the Serbian physicist Mileva Marič, and fathered three children, one out of wedlock, who was quietly given away. He separated from Mileva in 1914 and divorced her in 1919. As proven by their letters, Maric and Einstein discussed his theory although Maric (who had been Einstein's only female classmate) did not finish her degree or pursued a physics career.

3
The General Theory of Relativity

3.1 The Next Steps

In Newton's theory, gravity propagates at infinite speed. But the theory of special relativity says that nothing can move faster than the speed of light. There was therefore a contradiction between both theories. Einstein continued studying how Newton's theory should be modified—in particular, the gravitational force—to incorporate the results of special relativity. The problem was a hard one, and Einstein attacked it in phases.

The "Happiest Thought" of His Life

In a talk that Einstein gave in Kyoto on December 14, 1922, he described an episode that he recalled as taking place in 1907: "I was sitting in a chair in my patent office in Bern and the following occurred to me: if a man falls freely he would not feel his own weight. I was surprised that this simple imaginary experiment had in me such a big impact." He would later refer to this thought as the happiest in his life.

This example shows how an experiment—even an imaginary one—can lead to a theoretical conclusion. Galileo was perhaps the first modern physicist to imagine experiments. These *thought experiments* consist of extracting the fundamental aspects of a real and complex physical situation and "performing" the experiment with the imagination in a simplified situation. For instance, Galileo imagined a physical space without air friction to understand the fundamental aspects of the free fall of bodies. Einstein used thought experiments like Galileo did: to figure out the theoretical principle, it was not necessary to jump with a scale glued to one's feet from an airplane. Being able to think about the experiment and to extract the correct scientific conclusions from it is an act of intellectual bravery.

Precisely these thought experiments were crucial for the development of the general theory of relativity. In 1907, Einstein submitted a review article about the theory for publication in the journal *Yearbook of Radioactivity and Electronics*. In this article, he laid out three themes that he considered important and that were decisive for moving forward on the road toward the general theory: the principle of equivalence, the gravitational redshift, and the deflection of light.

The Principle of Equivalence

The principle of equivalence lies at the foundations of the general theory of relativity. It basically states—along the lines of Einstein's "happiest thought"—that a system with constant acceleration is indistinguishable from one in a uniform gravitational field. We have just introduced for the first time the idea of a field, a very important concept in physics. The reader may have heard about "electromagnetic fields" that use the same idea. A *field* is a quantity that is defined at every point in space and time (space-time!). For example, wind can be described as a velocity of air at every point, and then it is a field. For an electric field, we define it as the electric force that would act on an electric charge at every point, even if there is no actual electric charge (the field is produced by other electric charges). Of course, if we put an electric charge at that point, the force on it will be determined by the field we calculated. Similarly, the gravitational field produced by a star or a planet (or any mass) is the gravitational force it would produce on a mass at some distance from it, even if there is no mass at that distance.

A *uniform field* is one that has the same magnitude and direction everywhere. When we are on the surface of the Earth (a state in which we spend pretty much all our lives), we feel the gravitational field of the Earth. It varies little around us, because it depends on the distance to the center of the Earth, which does not change too much even if we move to the top of a mountain. So we can consider it for all practical purposes to be a uniform gravitational field.

Einstein's happiest thought can be understood with a simple experiment. Let's imagine a rocket that travels to outer space, far away from the Earth or any planet or stellar object, where the gravitational field has a negligible magnitude. Moreover, the rocket is moving with an acceleration equal to the one that an object falling near the surface of Earth would have (the acceleration Galileo measured). If a person weighed herself in a scale on the floor of the rocket, she would measure the same "weight" as on Earth: that is, she could not distinguish between whether she is in an accelerated rocket in outer space or in a rocket parked on Earth if there are no windows in the rocket (see figure 3.1).

The Gravitational Redshift

"Redshift" is the name given to the lowering of frequencies of waves (this is because the waves that constitute visible light become redder if their frequency becomes lower).

We experience shifts in frequencies due to the relative motion of objects: for instance, the change in the sound frequency of an ambulance or a firetruck siren as it moves away or toward us can be clearly perceived by our ears. When the ambulance is approaching us, the sound waves emitted by the vehicle are reaching us more often than when emitted, so the frequency (pitch) is higher. When

Figure 3.1
Acceleration at the Earth and in a rocket. A person weighing herself on a scale inside a rocket in outer space moving with acceleration equal to Earth's gravity (g) weighs the same as she would on Earth.

the ambulance recedes, the pitch is lower, because the waves reach us less frequently.

A less familiar example is police radar. The police check the speed of cars using a transmitter of electromagnetic radiation (usually microwaves or radio waves) combined with a receiver. When this device is pointed at a car, it sends an electromagnetic signal that bounces off the vehicle and is reflected back to the receiver. The device compares the frequency of both the emitted wave as well as the reflected one. If the car is not moving, both frequencies are the same. But if the the car is moving, the reflected wave frequency changes, and the change is a measure of the car's velocity: this is the operating principle of Doppler radar and the reason for many speeding tickets.

What does redshift have to do with gravity? Einstein wanted to extend special relativity to accelerated systems. He was particularly interested in accelerated systems in which the speed changes at a constant rate, like it does for the acceleration of gravity. Let us think about two observers at some distance from each other who are accelerated in the same direction: we can even think that both have the same (constant) acceleration, which is equal to the acceleration of gravity. If the observer who is behind sends an electromagnetic signal to the other observer (turning on a flashlight, for example), she will receive the signal after some time, but then her velocity will have changed (because she is accelerating), so there is a relative velocity between the emitter (the flashlight) and the observer. This means that the signal received will be shifted to lower frequencies: it will be redshifted.

Einstein concluded that observers who are accelerated would measure a shift in the frequency of the light. But according to the principle of equivalence, accelerated systems of reference are indistinguishable from a gravitational field. From this, Einstein concluded that the light will shift its frequency in the presence of a gravitational field, like that of the Earth.

The light changes frequency, shifting more to the red, the higher the acceleration or the more intense the gravitational field. Robert Pound and his graduate student Glen A. Rebka, Jr., performed an experiment verifying the gravitational redshift prediction in 1959 (again, decades after the prediction). The experiment consisted of using a gamma ray source emitting from the top of a 22 m tower and measuring the change of its frequency with a receiver positioned at its bottom.

The Deflection of Light

Einstein predicted another consequence of the principle of equivalence for effects of gravity on light rays, in addition to the redshift. In the presence of a gravitational field, light does not travel in a straight line. Its path bends. This had been already considered by other scientists, assuming that gravity acts on light beams like it does on massive objects. In 1801, Johann Georg von Soldner, a German physicist, mathematician, and astronomer, published an article titled "On the deflection of a light ray from its rectilinear motion," calculating what we call now Newtonian bending of light. The light deflection produced by the Sun would be about 0.8 arc seconds as seen from the Earth (an arc second is 1/3600 of 1 degree). This is for a ray that just misses the Sun (the effect is larger the closer to the Sun the ray passes). In 1911, Einstein revisited the effect and expressed strong interest in an experimental proof of this effect:

> It is greatly to be desired that astronomers take up the question broached here, even if the considerations here presented may appear insufficiently substantiated or even adventurous. Because apart from any theory, we must ask ourselves whether an influence of gravitational fields on the propagation of light can be detected with currently available instruments.[1]

In section 3.3 we will discuss such experimental attempts. Remarkably, the correct calculation in the general theory of relativity of the effect (not available in 1911) predicts twice the magnitude for it, due to the curvature of space.

3.2   Physics Calls Geometry for Help

After the early approaches to a relativistic theory of gravity, Einstein did not work much on the subject until 1911, when he took a job as professor at the University of Prague. Einstein's papers published in 1907 assumed space was still considered flat: the familiar space that scientists call "Euclidean," where parallel lines never cross. By 1912, Einstein convinced himself that he needed to search for a new gravitational dynamics, and he understood that the laws of Euclidean geometry that the Greeks had developed and had been used for 2,000 years were no good for this purpose.

Paul Ehrenfest was an Austro-Hungarian physicist who spent a great part of his career at the University of Leiden, in the Netherlands and had met Einstein in 1910 in Prague. Ehrenfest had formulated a paradox that greatly influenced Einstein. In a thought experiment, he applied the principles of the special theory of relativity to a rotating frame of reference. Recall the discussion in section 2.3 of our couple María and José experiencing the Galilean addition of velocities. José and María are examples of observers in inertial systems of reference—that is, observers who are only differentiated from each other by a constant velocity. The Ehrenfest experiment also involves two systems of reference, for which we can also think of María and José. But in this case, José is standing on a spinning disk, and María is watching him from a point outside the disk. The spinning disk is not an inertial system of reference. It is an accelerated one, because all points on the disk, like the ones along the rim, are changing the direction of their velocity: they are experiencing a centripetal acceleration. But a result from special relativity states that the times and lengths that observers in relative motion measure are different. María would perceive that the circumference of the disk is shorter than the one measured by José (see figure 3.2). And at the same time, she would not perceive a change in the radius of the disk because, being perpendicular to the direction of motion, it does not experience any shortening from her perspective. The effective result is that María would measure a circumference that is not equal to the number π multiplied by the disk diameter, which is the standard result from Euclidean geometry. But this means that the geometry involved is no longer Euclidean; it is one where the space is not flat but curved or warped.

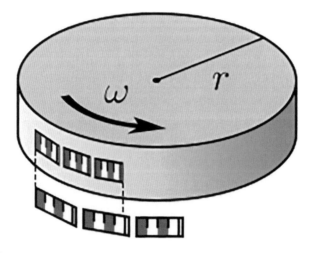

Figure 3.2
The length of the circumference of the rotating disk contracts for an observer external
to it, but the radius does not. (Credit: © 2013 Geek3/GNU-FDL, commons.wikimedia
.org/wiki/File:VFPt cylindrical magnet thumb.svg)

The logical consequence of this paradox is that in accelerated
frames of reference, the geometry is not flat anymore. And due to
the equivalence between accelerated systems and gravitation, this
means that a gravitational field must be associated with a non-flat
geometry. This paradox suggested to Einstein that a new geometry
was needed and that it would play a crucial role in his theory. The
theory of special relativity included new geometric ideas. Observers
who carry out experiments in systems in relative motion measure
different times and distances. The Lorentz transformations provide
a way to consistently describe the motion of a given body from two
different systems that are in motion with respect to each other. The
fact that Lorentz transformations involve both space and time put
them on an equal footing, without time having a preferred role.
This led to the concept of a space-time continuum in four dimensions (three spatial ones plus time as a fourth dimension). This geometrical construction received the name "Minkowski space-time."
But this new geometrical construction still had a Euclidean or "flat"

nature. To naturally incorporate gravity, a different type of geometry, a "curved" geometry, was needed.

Non-Euclidean Geometries

Euclidean geometry is quite intuitive. Euclid formulated the principles of geometry approximately three centuries before the Christian era. Little is known about him, but the Arab mathematicians transmitted his work to subsequent generations. Of particular notice is *The Elements*, which is the first mathematical compendium in history and very likely it condensed the work of many mathematicians prior to Euclid. The formulation of postulates and the proofs of theorems described in that book are still at the foundations of mathematics today. One of those postulates states that parallel line never cross. One of its consequences is that the interior angles of a triangle always add up to 180 degrees.

Starting with Ptolemy, many mathematicians tried to prove that postulate, but they did not succeed. A simple example shows that the postulate does not apply to curved surfaces like a sphere: figure 3.3 shows a triangle in which the sum of the interior angles is larger than 180 degrees. Also on the sphere, we see that parallel lines (meridians) do cross each other (at the north and south poles).

In 1854, the German mathematician Bernhard Riemann defined non-Euclidean geometries, which today we call "Riemannian geometries." The great merit of this formulation is that it prescribes how to measure distances, how to define parallels, and how to identify curvature in an unambiguous fashion (as in the case of a spherical surface, for example).

To describe geometries in a practical sense, we think about measurements of distances using rulers, or coordinates. In two dimensions, we can imagine measuring distances with lines on a grid paper: distances between corners are all the same, measured left to right and up to down. We can measure distances on straight or

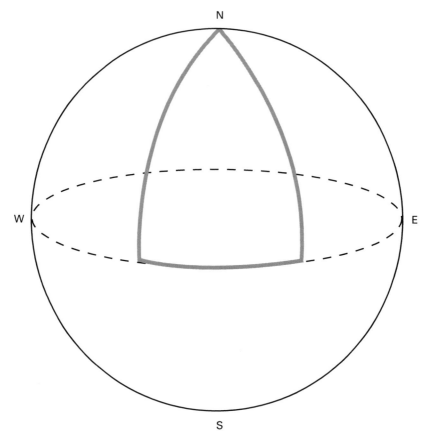

Figure 3.3
A triangle on a sphere showing that the sum of its interior angles is larger than 180 degrees: the two angles on the equator W-E already add up to 180 degrees, because they are right angles.

curved lines by counting distances between corners (dividing the squares into smaller ones if we need to). We can also measure the area of a figure, counting the squares within the figure. The lines we use for these measurements are called "coordinates" or a "coordinate system." If we rotate the grid paper 45 degrees, we can still use the lines to measure distances and areas, and we would get the same values for the same curved lines or the same figures as before. The rotation of the grid paper is called a "change of coordinates," and the fact that the measurements of areas and lengths are the same in both systems means those quantities are invariant, or independent of the coordinate system. Of course, the real world has three spatial dimensions, so we need a three-dimensional coordinate system; moreover, Einstein showed that time was another dimension. That four-dimensional space-time can be visualized as a three-dimensional grid with clocks on each corner of the grid like the one shown in figure 3.4. In a flat, or Euclidean, space-time, all distances between corners are the same and all clocks are synchronized with each other. In curved geometries, strange things happen.

Einstein Integrates Physics and Geometry

Einstein formulated the theory of special relativity by defining inertial frames in which the laws of physics are the same and the speed of light has the same value in all of them. The "laws of physics" he considered were Newton's law of inertia (if there are no forces, a body moves with constant velocity) and Maxwell's laws of

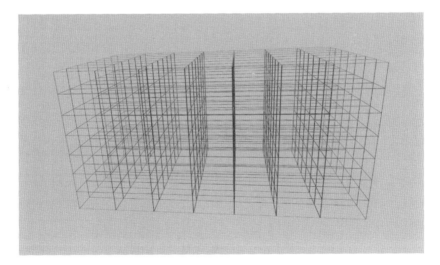

Figure 3.4
Artist's conception of a flat space-time as a three-dimensional grid. (Credit: Américo Hinojosa Lee.)

electromagnetism. The inertial frames Einstein considered moved with constant velocity with respect to each other: distances and times in one system may be different than in another inertial frame (due to Lorentz contraction and time dilation), but all distances between corners in one system are the same, and all clocks in corners of the same system are still synchronized.

Following his work on the formulation of the equivalence principle, the gravitational redshift, and the deflection of light by gravitational bodies, Einstein was ready to take the next step: gravity. He extended the two principles of special relativity to non-inertial (accelerated) systems as well. Now, following his "happiest thought," he was closer to incorporating gravity into special relativity.

Einstein needed a formulation that did not change the laws of motion when the space-time coordinate systems are changed. But these coordinate transformations were now much more general than those of special relativity: they included accelerated frames.

Consider the airport moving walkway that we discussed in chapter 2. Assume now that the walkway is moving at a changing speed— that is, accelerated—with the same acceleration as that of gravity (that is a lot of acceleration: it would violate safety rules at airports, but this is a thought experiment). Walking at a constant speed on the walkway, Alice is going to feel like she is in a gravitational field that pushes her backward. Nevertheless, the laws of physics must be the same for her and for Bob, her companion, who is still walking next to the walkway on the airport floor. How are the coordinates of two different systems related to each other so that the laws of physics remain the same? Einstein's first step consisted of constructing a theory that leaves invariant a particular mathematical quantity: the distance computed using a tool called the "metric."[2]

The tool yields a formula that we can use to calculate the shortest distance between two points. Whereas the shortest distance in the Euclidean space is given via a segment of straight line that connects both points, in a curved space—for instance, on a sphere—this would make no sense: a straight line would just leave the surface. The metric allows us to compute precisely that shortest distance in any type of geometry; for instance, on a sphere, the distance between two points is obtained by the length of the arc of the maximum circle that passes through them.[3] For more complicated geometries the math details exceed the scope of this book. The challenge was so huge that—according to some historians—Einstein asked a friend and classmate for help, the Swiss mathematician Marcel Grossman. With his help, Einstein understood that the key to finding a system of coordinates in which the gravitational field (the acceleration) disappears consisted of using the formulation that had been developed in the nineteenth century by Riemann.

Einstein was also sensitive to the fact that the new theory should yield the same results as Newton's theory in the situations where the earlier theory had already proven itself successful, such as describing the motion of planets, which move at low velocities compared to the speed of light. Einstein dedicated 8 years of patient effort and hard work, and he finally was able to develop a theory that generalized special relativity: "general" relativity. In such a theory, the laws of physics were the same, even if one changed to a system of reference that was accelerated. The theory also incorporated the effects of gravity as a phenomenon equivalent to the acceleration of a reference frame (recall his happiest thought, the example of the rocket). Moreover, Newton's laws of gravity were a prediction of the theory as a special case when gravity is weak and does not change too much over time.

Einstein introduced his theory to the Prussian Academy of Sciences in four papers in November 1915. In this theory, space is not a mere frame of reference or a stage on which physical phenomena take place. Space-time—its geometry—is determined by the matter and energy in it: massive bodies and the way they move are the reason that space-time is curved, as shown in figure 3.5.

The Einstein equations are formally so simple and beautiful that we decided that it would be one of the few formulas we would

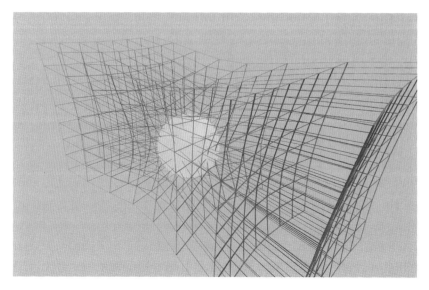

Figure 3.5
Artists' conception of the curved space-time due to the presence of a spherical mass.
(Credit: Américo Hinojosa Lee.)

display in this book (up to now, we only have included $E = mc^2$), and even if readers are not mathematical experts, we are sure that you will be able to appreciate their conceptual simplicity:

$$G = kT.$$

This is the relativistic theory of gravity (just like special relativity describes relativistic mechanics). The term G is called the "Einstein tensor."

A tensor is a generalization of another geometrical construction: a vector. A vector in three dimensions is a collection of three numbers that identifies in a unique way physical or geometrical quantities. For example, a point in space can be given by three numbers: the distances from a given origin of coordinates (left-right, back-forth, and up-down). The velocity of an object is also a vector: the three numbers define a unique direction and magnitude for it. There are situations in which more numbers are needed, for example, if we want to describe how an extended body is stretched or squeezed, we need another type or array of numbers called "tensors." The Einstein tensor is such an array of functions (ten of them!) of the coordinates of space-time. The tensor G is laboriously calculated from the metric: G charts the geometrical structure of space-time. The constant k is a combination of numbers, Newtons' constant of gravity, and the speed of light. On the right-hand side of the equation, the quantity T is also a tensor that contains all the information about matter and energy that exist in the space-time (including the motion of that matter). These equations relate the geometry of space-time to its matter and energy content. The relationship is technically what it is known as a nonlinear one. The matter content (described by T) conditions the geometry (itself expressed by G), and the geometry tells the matter how it should move.

In spite of their conceptual simplicity, we have to warn the reader that mathematically, Einstein's equations are very complex. If the matter content (the T tensor) is known, then the equations allow us to figure out the curvature of the space-time (from the tensor G): this is called "finding a solution" to Einstein's equations.

In a flat space-time, the shortest distance between two points is given by a straight line. In a curved space-time, the shortest distance between two points is given by a curve known as a "geodesic." For example, on a sphere, the geodesics are maximum circles. In Einstein's general relativity, particles whose masses are not significant enough to affect the gravitational field created by all other matter, move along geodesics. Thus, the Moon goes around the Earth not because there is a force attracting it, but because the Earth curves space-time in its surroundings, and an object moving through it does not go in a straight line: it follows a curved trajectory that corresponds to the orbit of the Moon. And so the theory explains the motion of the planets as Newton's laws did (in a mathematically much more complicated way), but with higher precision and better predictions. This is particularly so when the fields are stronger (like the gravitational pull experienced by

Mercury) and the speeds involved are much higher than the ones in most common situations.

3.3   The Early Successes of the New Theory

The new theory predicted new physical behaviors. Some were verified reasonably quickly, others took longer to be understood and confirmed. Here we consider the former, which first made Einstein famous.

Mercury's Perihelion Precession

An immediate impact the theory had in the scientific community was to explain a great astronomical mystery of the time: the precession of Mercury's perihelion.

Let us describe what this is about. The point of the orbit of a planet that is closest to the Sun is called the "perihelion." This point slowly rotates, and for Mercury, it exhibits an accumulated variation that became significant enough to be measured after a century. This measurement disagreed by a bit less than 1 percent with the predictions based on Newton's theory.

Before Einstein, many possible explanations were attempted: maybe the mass of Venus was larger than it was thought to be (this method was used to predict the existence of Neptune before it was discovered), or maybe an unknown planet was present between the Sun and Mercury. Or it could be that there was a swarm of asteroids between the Sun and Mercury perturbing its orbit. It was even speculated that Mercury could have a moon. But none of these hypothetical solutions had been observed. Einstein used his equations to compute the orbit of Mercury, and he felt a great joy when he discovered that his calculations yielded a value for the precession of the perihelion equal to the one observed by astronomers.

The Deflection of Light

The new theory also established that light did not travel in a straight line. In particular, it predicted that the trajectory of light coming from a star and passing near the Sun should deviate from a straight line. As we mentioned, Einstein had already calculated this effect using special relativity in 1911, agreeing with earlier calculations using Newtonian mechanics. Although the effect predicted by general relativity was twice as large, it was still very small and difficult to measure. It is impossible to observe directly the light of a star that is near the Sun in the sky, since stars are not visible in daylight. The direct observation, however, is possible during a total solar eclipse, when the Moon blocks the sunlight. A comparison between the position of the star on a photographic plate taken during an eclipse and the position of the same star when it is visible during the night—without the light traveling close to the Sun—allows verification of the theory's prediction. The light of the star deviates: that is, the star would appear "displaced" with respect to its nightly position when the light trajectory is not perturbed by the Sun.

Obviously, to predict correctly a new effect is a bigger success for a theory than simply to reproduce already known results. This effect had not been measured before the introduction of Einstein's theory, even though there had been some attempts following his call to astronomers in 1911 to do so. One such astronomer from Berlin was Erwin Freundlich. He in turn was friends with the director of the Astronomical Observatory of Córdoba in Argentina, the American astronomer Charles Perrine: as a consequence, the first expedition to try to measure the deviation of light by the Sun was Argentinian, like the authors of this book.

Several astronomers from Córdoba went to Cristina—some 200 kilometers northeast of São Paulo in Brazil—to attempt the measurement during the eclipse of October 10, 1912. Unfortunately, it was cloudy during the eclipse, and the measurement could not be carried out as the stars were not visible. The Argentinians lent their equipment to Freundlich, who attempted the measurement in 1914 in Crimea, but the outbreak of World War I interrupted the project. A third Argentinian attempt took place in Tucacas, Venezuela, in 1916, but the limited equipment they took and the bad general conditions dissuaded them from attempting the measurement; they instead concentrated on other aspects of the eclipse.

Due to budget limitations, the Argentinians could not travel to perform the experiment during the eclipse of 1919, visible in Sobral, Brazil, and also on the island of Príncipe, off the African coast. This eclipse was observed by a British expedition led by (the future Sir) Arthur Eddington. His observations confirmed the prediction of general relativity, which launched Einstein to stardom, becoming the first example of a "scientist-celebrity." It is interesting to speculate about what would have happened if the measurement had taken place before general relativity, when Einstein's calculations seemed to agree with Newton.

The *New York Times* of November 10, 1919, published Eddington's observations with a weird headline: "LIGHTS ALL ASKEW IN THE HEAVENS; Men of Science More or Less Agog Over Results of Eclipse Observations. EINSTEIN THEORY TRIUMPHS. Stars Not Where They Seemed or Were Calculated to Be, but Nobody Need Worry. A BOOK FOR 12 WISE MEN. No More in All the World Could Comprehend It, Said Einstein When His Daring Publishers Accepted It." The reason for this surreal announcement is that the only journalist *The New York Times* had in the United Kingdom at the time was a sports correspondent covering the British Open golf tournament. However, the myth that only a few people understand Einstein's theory can still be found on the internet today.

Eddington was a conscientious objector during World War I, and his refusal to be drafted was punished with a prison sentence. The Royal Astronomer had to intervene for Eddington not to go to prison, and partly as an alternative punishment, he was assigned "military duty" to command the expedition to Sobral and Príncipe, which were particularly inhospitable places at the time. Eddington admired the pacifist convictions of Einstein and considered the fact that a British scientist confirmed the theory of a German could provide a good gesture of postwar reconciliation.

Throughout the years, some questions have arisen regarding the possibility that Eddington had "fudged" the results. Several of the photographic plates had been discarded, and apparently they were among the best ones, but they did not confirm Einstein's theory. The idea that Eddington had selectively chosen the data was presented in the book *The Golem*, by Harry Collins and Trevor Pinch in 1993, but it is likely that we will never know what happened. A detailed study of all the available documentation carried out by Daniel Kennefick in his 2019 book *No Shadow of a Doubt* suggests that Eddington did not alter his results; in fact, apparently he did not himself carry out the plate selections. Today the deviation of light—and radio waves—by the Sun has been verified to a high level of precision using other techniques, and the results clearly confirm Einstein's theory.

3.4   The Shocking Predictions of the New Theory

In addition to predicting physical effects that were directly measured, as we described in the section 3.3, implicit in the theory were other physical results. The latter ended up being really spectacular, but they took considerably longer to verify experimentally.

Black Holes

Einstein's equations are considered equations for the metric (see section 3.2). Given a known distribution of matter and energy, we can calculate the tensor T, and we want to find a metric such that when we use it to calculate the tensor G, we get exactly kT: this would be an "exact" solution. The equations have parts of them (terms) that if velocities are small, and we are far from masses, and gravity is weak, we can disregard. This leads to approximate solutions. This is what Einstein did to calculate the Mercury's orbit around the Sun using his theory. However, he did not obtain the first exact solution.

Two months after having presented his theory in the Prussian Academy of Sciences, on January 16, 1916—during World War I—Einstein presented at the Academy an article on behalf of Karl Schwarzchild, a German astronomer who couldn't do it himself because he was deployed with the German Army on the Russian front during the war. On February 24, Einstein presented another article, also on Schwarzchild's behalf. Both articles described the first exact family of solutions of the Einstein equations, assuming that some amount of mass is not moving and that the metric will be the same at all points at the same distance from the coordinate origin (this is called "spherical symmetry").

One of the solutions described by Schwarzschild had two remarkable features: (1) the space-time is empty, except at the very center where all the mass is concentrated, and (2) the solution includes a sphere of a radius that depends on the mass, with very special properties. The distance from the center to the location of this surface is now called "Schwarzschild radius" for obvious reasons. It is also called the "event horizon"—you may have heard this term in the news in 2019 and in several Hollywood movies. We are talking about the same thing! We'll try to explain it carefully, though. Please be patient.

First, we define the concept of the escape velocity of a planet (or a star, or any spherical massive object). You are familiar with the statement "everything that rises will fall," and can verify it is true when you throw a stone up in the air. However, in general, this is not true: humans have sent spaceships far away from Earth, and some of them will never come back. For this to happen, the spaceship needs to rise from Earth faster than 11.2 km/sec (around 30,000 miles per hour): this is Earth's escape velocity. The latter depends on the mass and radius of the astronomical body it refers to. For a star like the Sun, an object on its surface (assuming it could take the tremendous heat) would require a speed of more that a million miles per hour to escape its gravitational attraction.

An object at a distance from a given body equal to its Schwarzschild radius would require an escape velocity equal to the speed of light. But as we have seen, this is a speed limit of nature: according to Einstein's theory of special relativity, nothing can go faster than light. Therefore, if an astronomical object has a radius smaller than its Schwarzschild radius, nothing can escape it. Nothing on this object could communicate with any other body beyond it, because neither light nor any other kind of matter or energy would be able to escape.

In general, the Schwarzschild radius is very small compared with the physical sizes of stars and planets. In the case of the Earth, the distance equivalent to the Schwarzschild radius is about 1/3 of an inch, and for the Sun it is around 2 miles. So objects need a large—but achievable—speed and can escape the Earth or the Sun. However, if the mass of the Sun were concentrated within those 2 miles, or the mass of the Earth within 1/3 of an inch, light could not escape. If the Sun were smaller than 2 miles in radius, the planets in the solar system would still move in the same orbits (because the mass of the Sun would be the same), but all planets would be dark and cold, revolving around a black star.

This is why the spherical surface at a distance from the center of the star equal to the Schwarzschild radius is known as the "event horizon." Any event that takes place inside that surface cannot influence the exterior. The Schwarzschild solution describes what much later came to be known as a "black hole." In fact, its meaning was not properly understood until the 1960s, as we will discuss in section 5.4. A black hole is not only invisible to telescopes because it doesn't emit light, but it is also surrounded by an event horizon without communication with the rest of the universe. Years later, other exact solutions to the Einstein equations were found that represent rotating and electrically charged black holes, but Einstein himself did not believe in the actual existence of these objects. In fact, he died in 1955 without understanding them properly. Objects with their mass concentrated at a single point made the density (mass per unit volume) infinite there: for him, this was all probably just a mathematical curiosity.

Although nothing can escape from a black hole, they are usually surrounded by matter that the strong gravitational fields heat up and make bright at various wavelengths. On April 10, 2019, the first image of a black hole was published by the Event Horizon Telescope Collaboration. After several years of observations, they managed to map, using several radiotelescopes all over the world, the event horizon surrounding a supermassive black hole at the center of the galaxy Messier 87. All the studies made of the data obtained verified Einstein's theory and the expected properties of the Schwarzschild solution.

The Big Bang

All religions and mythologies outline some kind of cosmogony; the origin and future of the universe has intrigued human beings probably since we started roaming the planet. Most humans resorted to divine intervention, or speculations of a metaphysical nature. Probably the first attempt to discover the origin of the universe and its nature using physical laws was that of the British theologian Robert Grosseteste. In his treatise written in 1225 titled *De Luce* (*About Light* in Latin), he ascribed the origin of the universe to a great explosion followed by the transformation of light into matter to form the stars and the planets, organized in concentric spheres centered on the Earth. Grosseteste's works and those of his disciple, Roger Bacon, precede and constitute the embryo of the scientific method, which Galileo Galilei would formalize about 200 years later. But only in the twentieth century, with the technological development that observational astronomy had acquired, were fundamental aspects of the structure of the universe actually verified.

Our understanding of the world was limited until the Renaissance, but during the 500 years since then, the methods and tools that would lead to the expansion of the economy and the development of scientific and technological knowledge were slowly developed. It was in the twentieth century that for the first time, speculations about the origin and formation of the universe arose based on direct observations, establishing the foundations of modern cosmology.

Milky Way, Via Láctea in Spanish, Voie Lactée in French, all derive from the Greek word "gala" meaning "milky." In the sixteenth century, Galileo—the first to use a telescope to look at the sky—observed that the Milky Way was composed of countless stars. During the eighteenth and nineteenth centuries, astronomers had observed "dots" of light that were not stars and called them "spiral nebulae." They thought these dots were clouds of gas or clusters of many stars and believed that these formations were contained in the Milky Way. During the nineteenth and beginning of the twentieth century, more evolved methods allowed a better survey of our galaxy. It was then established that it has the shape of a disk 100,000 light-years[4] in diameter, with "arms" forming a spiral, as shown in plate 2. In the Milky Way, the Sun is in one of the arms, removed from the center.

In the first 20 years of the twentieth century, a very different landscape emerged than what had been imagined before then: not only was the Earth not at the center of the universe, but the Sun—the star around which we orbit—was also very far from the center of our galaxy. And by then, many astronomers knew that the "spiral nebulae" were also galaxies like ours. Today astronomers estimate that the total number of galaxies in the universe is more than a few billion.

In 1917, shortly after making his equations public, Einstein noted that they predicted a dynamic universe that was expanding. But at the time, it was still thought that the universe was static: planets and stars could be moving, but the average density was the same when considered at large scales. Einstein then added a term to his equations to force the resulting model of the universe to be static. When it was verified that the universe is expanding, it is said that Einstein claimed that introducing the extra term was "the biggest blunder of my life."

The discovery of the expansion of the universe happened in the years close to Einstein's publication of general relativity and is an interesting story. In 1910, the American astronomer Vesto Slipher—followed later by Carl Wilhelm Wirtz in Germany—observed that light emitted from the spiral nebulae (today identified as spiral galaxies) is redshifted, which implies that they are receding from the Earth. Moreover, this phenomenon occurs for galaxies in all directions in the sky. Was the Earth again at the center of the Universe? These astronomers did not interpret their observations in this way, nor did they suspect the cosmological consequences of that discovery, which was realized a decade later: the universe is expanding.

In 1922, Russian mathematician and physicist Alexander Friedmann applied the Einstein equations to describe cosmological models, assuming for simplicity that the universe is isotropic (it looks the same in all directions) and homogeneous (the distribution of matter in space at any time is uniform). The solution that Friedmann found describes a universe that is expanding. Although today we consider that these solutions correctly describe our universe at large scales, Friedmann's work did not have much impact at the time.

In 1927, the Belgian Catholic priest Georges Lemaître proposed a cosmological model similar to Friedmann's that starts in what is mathematically called a "singularity" (a point where physical quantities become infinite). If the universe is expanding at present, then it must have been smaller in the past. In fact, if one runs the model back in time, at some instant in the past, all matter and radiation must have been concentrated at a single point. Such a singularity was later disparagingly called "Big Bang" by the astronomer Fred Hoyle (he did not believe in it). Today the term is part of everyday language, even becoming part of the name of a popular TV sitcom.

In 1929, the American astronomer Edwin Hubble discovered a relation between the velocity of recession of the galaxies—their movement as they get farther away from us—with the distance to them as measured from Earth: the farther the galaxies were, the faster they were receding. This experimental observation could be explained by Friedmann and Lemaître's ideas: the universe expands according to an exact law, which we now call "Hubble's–Lemaître's law."[5] This is a phenomenon that any observer in any galaxy could also observe.

The Big Bang theory has been successfully verified through countless observations during the past 80 years. According to this theory, the universe started about 13,800 million years ago with a big explosion. Before the term "Big Bang" was coined, Lemaître called the phenomenon the "cosmic egg." The conditions during the instant of the explosion are not known in detail. It is highly likely that a violent expansion (called "inflation") increased the volume of the universe enormously in a brief instant of time after the Big Bang. During those first instants, the first elementary particles were created that later would form protons, neutrons, and electrons. In the first 3 minutes after the great explosion, the first atomic nuclei were formed from these particles. But the electrically neutral atoms that constitute all known matter, including the Sun, the Earth, and ourselves, took several thousand years to form.

The tiny fluctuations originating in the "cosmic egg" during the gigantic explosion constitute the seeds that allowed the atoms and molecules created—fundamentally hydrogen and in lesser quantity helium—to be distributed and clumped together with certain irregularities. The generated clumps gave origin to the stars and galaxies, and eventually to all other elements of the periodic table. The combination of the Einstein equations with the physics that studies atoms and nuclei (nuclear physics) has very successfully explained the abundance of the known chemical elements. All these predictions can be seen as another triumph of Einstein's theory.

4

Gravitational Waves: Their Long and Difficult History

For the first half-century after its publication, the general theory of relativity attracted only theoretical physicists to work on gravitation. After all—as Richard Feynman put it—it was a field with no experiments. After the measurement of the deflection of the light from a star during an eclipse and the explanation of the displacement of Mercury's orbit, there were no new experiments to compare with predictions for nearly 50 years. Only the development of astrophysics, which accelerated after World War II, would lead to a series of discoveries that required general relativity for their explanation.

For a long time, progress in the general theory of relativity was slow and convoluted. Einstein's equations pose a mathematical problem that is very difficult to solve. For example, in the formulation of gravity by Newton, the so-called "two-body problem"—like the motion of the Earth around the Sun—has an exact solution (the Earth moves in an ellipse around the Sun). The same problem does not have an exact solution in general relativity. But this does not mean that approximate solutions cannot be found having a high level of precision, using very powerful supercomputers, as we describe in chapter 7.

A complete understanding of one of the more controversial predictions of the theory—gravitational waves—had to wait for investigations that took place in the 1970s and later.

4.1 The Prophecy

How did the theory predict gravitational waves?

Physical theories are formulated in terms of equations. They involve known quantities, which we call "constants," and other, unknown, quantities we want to find, which we call "variables." A simple example is the equation $5 = x + 3$. In this case, 5 and 3 are constants, and x is the variable. The solution of the equation is the value of the unknown variable, in this case $x = 2$. The example given is what is known as an *algebraic equation*: its solution is a number. There exist more complicated equations; they involve, in addition to constants, certain operations that act on the variables, like derivatives. Derivatives are mathematical operations that measure how much something changes: for example, the position of a little ant crawling at a constant pace of 1 cm/sec on a wire changes with time; we call this a "function": $x(t) = 1$ cm/sec$\times t$ sec. The derivative of the position as a function of time in this case will give the speed of the ant: $dx(t)/dt = 1$ cm/sec. These types of equations determine how much a quantity differs from a previous value and hence they are known as *differential equations*. Newton invented what we now call "differential calculus" for developing his theory of gravity. All the equations in physics (like Einstein's) are differential equations.

The constants of the differential equations depend on the situation being considered. For example, in Newton's theory, we can consider two, three, or more masses. We can also consider the masses to be spherical, which even if not completely realistic, simplifies the equations and is approximately true for the Sun and the planets.

To obtain exact solutions of the Einstein equations is very difficult (although the three authors of this book started their careers doing just that!), and it is only possible in very simple cases. As mentioned, the first exact solution found by Schwarzschild was done assuming the mass was spherically distributed.

Einstein's equations are nonlinear. A linear relation can be exemplified by the distance traveled by a car moving at a constant speed: double the amount of time, and the car will travel twice the original distance. An example of a nonlinear relation is the case in which the same car accelerates at a constant acceleration: its speed changes linearly with time but in this case, the distance changes quadratically with time—that is, time squared—. In three times the time, it will travel nine times the distance.

One way of solving Einstein's equations $G = kT$ (and many other equations) is to find solutions by starting from a previously known one, which typically represents a simple case, assuming the real solution differs only slightly from the simple one. This allows us to simplify considerably the equations, disregarding the nonlinear terms, because they become very small (a small quantity squared results in an even smaller quantity). The result is an approximate solution, not an exact one. This method is known as linearization.

The first solutions that Einstein investigated consisted of modeling situations in which the starting known solution is flat space-time (i.e., the Minkowski space-time), a solution when the stress energy tensor is zero, $T = 0$. There is no matter, and we call this a "vacuum equation," where the Einstein tensor is also zero: $G = 0$. The variable in Einstein's equation is not G, but it is the metric we use to measure distances and times. As we mentioned, the tensor G involves derivatives and products of the metric. All the terms in the Minkowski metric are constant, so its derivatives are zero, which makes the Einstein tensor zero. Was this the only solution? Einstein's idea was to find approximate solutions to $G = 0$, using the linearization technique and starting from Minkowski metric. Following this method—*the weak field method*—in 1916, he found solutions to his equations that represented traveling waves: gravitational waves.

4.2   Waves in Physics

Waves are a familiar concept, and they are a fundamental concept in physics. For example, a rope held by two ends under tension—as in the case of the strings of a violin—when plucked will undergo a vibration that propagates along its length (the string goes up and down or side to side). The vibration travels along the string in one dimension (as in shown in figure 4.1).

But there also exist waves that propagate in more than one dimension: for instance, the membrane of a drum, or the water in a lake in which a stone is dropped; these are examples of two-dimensional waves, because they can travel in two independent directions. Repetitive changes in air pressure make waves traveling through it, known as sound, which are three dimensional. Three-dimensional

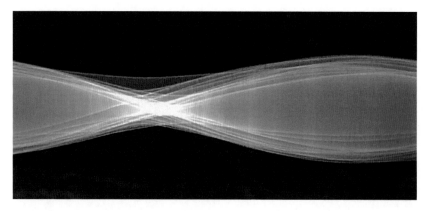

Figure 4.1
High-speed photograph of a vibrating string. (Credit: Photo courtesy of Andrew Davidhazy.)

waves can travel in three independent directions (up/down, right/left, forward/back). The motion of the particles in sound or water or drum or string waves are all solutions of the same equation: the wave equation.

As we discussed, in 1865, Maxwell formulated the laws of magnetism and electricity as four equations that implied the existence of electromagnetic waves traveling at the speed of light. And then there was light!

4.3   Gravitational Waves

As an interesting historical aside, the first attempt to understand gravity as relating geometry with the movement of masses (and even beyond that—that its action manifests as waves) was not done by Einstein. The idea was ventured by a young English mathematician: William Kingdon Clifford, who passed away at 34 in 1879 after a short but very productive life. Using the geometry known at the time, he investigated and understood very well the relationship between geometry and algebra. He also distinguished himself for his philosophical contributions, confronting the society of his time with his antireligious convictions. Clifford considered immoral the belief in theories or principles that were not supported by clear evidence. In a conference at the Cambridge Philosophical Society in 1870, after discussing the properties of Riemann's geometry, he declared that it could be applied to physical phenomena. He proposed that space had little "hills" on a flat surface and that the curvature passes continuously from one region to another "in the manner of a wave." Clifford was essentially talking about gravitational waves without using those words. His thought is a clear example of a theory ahead of its time. It would take another 46 years for similar ideas to attain maturity. And even after the mathematical discovery that Einstein's equations contain wave-like solutions, many developments would take place, including retractions, doubts, and confusion, until it was clarified what their sources were and how they could be detected.

The term *gravitational waves* was first used in July 1905 by the French mathematician Henri Poincaré, in an article about Einstein's special relativity. He noted that since nothing could go faster than light, then neither could gravity. If gravity takes some time to propagate, the process of its propagation leads to the existence of waves, disturbances that travel as time evolves. Einstein mentions gravitational waves for the first time in a letter we already mentioned to German astronomer Karl Schwarzschild in 1916. Schwarzschild had asked Einstein whether he believed that his theory predicted gravitational waves. And this is how the twisted and meandering history of the field starts. Einstein, without too much explanation, replied that the theory does *not* predict gravitational waves. But only 3 months later—in June 1916—Einstein published a paper showing the existence of gravitational waves using the weak field method.

The calculations were not straightforward. In particular, Einstein decided to use a certain coordinate system that he thought simplified calculations. In hindsight, his choice made things harder. Due to the complexity of the calculations, he confused two variables, and his final formula for how gravitational waves are produced was completely wrong. The mistake was noticed independently by the Finnish physicist Gunnar Nordström and the Dutch astronomer Wilhelm de Sitter. They both wrote to Einstein. He was forced to write a second article in 1918 correcting his errors. The 1918 paper is essentially correct (it misses a factor of 2) and presents a description of gravitational waves as one would find in a modern textbook. In particular, it includes the formula that shows how gravitational waves are produced, and he also concludes that there exist different types of gravitational waves.

In 1923, the English astronomer Arthur Eddington published a book on relativity. It was very influential, because it was the first description of the theory in English. In it, Eddington discusses gravitational waves. He corrects the factor of 2 that Einstein had missed. He also noted there were several types of gravitational waves and that two of them were truly of a physical nature, but the others were a mathematical artifact. If one changed coordinates, these waves disappeared. Eddington made the caustic remark that those specific gravitational waves "travel at the speed of thought." This phrase was misquoted by many physicists throughout the years as implying that all types of gravitational waves do not exist.

In the 1920s, the rise of anti-Semitism in Germany led, among other things, to a campaign against "Jewish physics," attempting to show that all physics results obtained by Jews were wrong. Einstein appeared in the crosshairs of this campaign. The subject of gravitational waves and its unclear standing in the community was included among the criticisms of Einstein's theory. In 1931, a book was published called *100 Authors against Einstein*. The well-respected German philosopher of science Hans Reichenbach has described the book as an "accumulation of naive errors" and as "unintentionally funny." When asked about it, Einstein himself replied "if I were wrong, a single author would have been enough."

4.4   The Sources of Gravitational Waves

Einstein found that the sources of gravitational waves are accelerated nonspherical distributions of matter. A perfectly spherical star that is spinning does not produce gravitational waves. Stars are not perfectly spherical, but black holes are: spinning black holes do not produce gravitational waves.

The measure of the nonspherical distribution of matter is its *quadrupole moment*.[1] Some examples of accelerated bodies with a nonzero quadrupole moment are a nonspherical spinning star, a planet going around a star, or two stars orbiting each other (a binary star system).

For a system to generate gravitational waves, it (1) must have a nonspherical distribution of matter; (b) this nonspherical distribution must change with time; and (c) the change must not be constant but must accelerate. In nature, these conditions are more common than it might at first appear: all stars rotate, and even if the rotation speed is constant, the velocity of the particles that make up the star is changing direction all the time. Thus the particles are accelerated (this acceleration is often called "centripetal acceleration"). The stars in a binary system are also accelerated: the direction of the velocity of each star is changing. But even if it rotates, a spherical star (which the Sun is, approximately) does not produce gravitational waves, because its quadrupole moment vanishes. However, two stars rotating around each other (as shown in figure 4.2) form a nonspherical physical system that is accelerated. The stars may be spherical, but the system is

Figure 4.2
A binary system of white dwarf stars. It generates gravitational waves of low frequency. (Credit: NASA/T. Strohmayer (GSFC)/D. Berry (Chandra X-Ray Obervatory).)

clearly not the same in the up/down direction as in the side-to-side direction. This system has an accelerated quadrupole moment and therefore produces gravitational waves.

Other examples are an exploding star like a supernova, if the explosion is not spherical. Or a nonspherical rotating star that has a mountain on it, as shown in figure 4.3. In this case, the physical system produces continuous gravitational waves in the sense that they do not pulsate, they continue indefinitely as long as the star and the mountain exist. The frequency of the gravitational wave is double the rotation frequency of the star. For a neutron star, the diameter is typically 10 km, and the mountains have a height of 3 mm!

In his 1916 paper, after deriving the expression for the strength of the gravitational waves produced by a source, Einstein said it

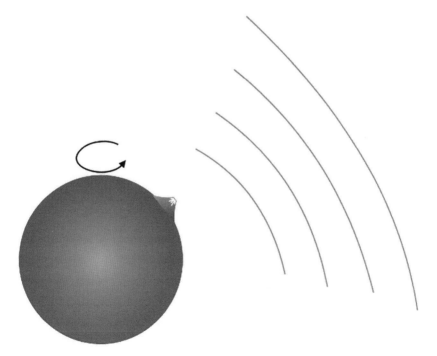

Figure 4.3
An asymmetric neutron star; the asymmetry comes from the presence of a small mountain on it.

"has, in all imaginable cases, a practically vanishing value." He did not elaborate on the possibility of observing them. It may appear surprising that Einstein did not attempt to estimate the intensity of gravitational waves of, for instance, the abovementioned case of two stars orbiting each other. But at the beginning of the twentieth century, knowledge of the universe was very limited. Little was known about stellar evolution (i.e., the life and death of stars), and no one at the time could imagine processes that would produce compact objects—massive and of small volume—capable of producing gravitational waves that were detectable with the technology of the time. It is possible that Einstein worked out the radiation produced by two stars and realized that it would have been completely undetectable with the technology of 1916 and so decided to omit his result from the paper. Even with today's technology, it is not possible to detect gravitational waves from the kind of stars orbiting each other that were known in 1916.

Only in 1941 did two Soviet theoretical physicists, Lev Landau (future Nobel Prize winner) and Evgeny Lifshitz, propose for the first time that the linearized calculations of Einstein could be applied to a binary system of stars, using the solutions found by Newton for their motion. The proposal appears in an exercise at the end of a chapter in a textbook they wrote and left to the reader to work through. (The exercises in this textbook are both feared and enjoyed by physics graduate students.)

The first published estimate of the intensity of the gravitational waves produced by two stars orbiting each other in a binary system was presented by the British–American physicist Freeman Dyson. Curiously, the computation was made in an article in a book titled *Interstellar Communications*, about how to communicate with extraterrestrials. He argued that gravitational waves could be evidence of great machines built by extraterrestrial civilizations.

4.5   Einstein's Doubts about Gravitational Waves

When the Nazis came to power in Germany, Einstein emigrated to the US. His arrival in 1932 was almost frustrated, because the Woman Patriot Corporation, a group that identified itself as anticommunist, anti-pacifist, and anti-feminist, wrote to the American consul in Berlin requesting him to deny Einstein a visa based on his pacifist and socialist ideas. Einstein was called to an interview and became irate when the vice consul started asking about his politics. He threatened to cancel his trip to the US and go public if he did not get his visa. He did call the *New York Times*, which printed a story on the incident. This brought pressure on the consul, who issued the visa. But the letter he had received was forwarded to the FBI, which started a file on Einstein. The file would eventually grow to 2,500 pages. J. Edgar Hoover's FBI tracked Einstein for the rest of his life, making several attempts to demonstrate that he collaborated with the Soviet Union. Part of the strategy was to pressure Einstein to self-deport. It is worthwhile noting that they pursued a similar strategy with the famous actor Charlie Chaplin—a good friend of Einstein—in his case successfully. The difference between Chaplin and Einstein was that the latter had become a US citizen. A book by Fred Jerome titled *The Einstein File* details this story.

Einstein settled in at the recently founded Institute for Advanced Study, a private institute located in Princeton, New Jersey, near Princeton University. There he worked with a pair of assistants, Nathan Rosen and Boris Podolsky. In 1935, they wrote a paper about quantum mechanics that became quite relevant with the advent of quantum technologies and quantum computing, becoming Einstein's most frequently cited paper, with most of the citations appearing in the twenty-first century. In 1936, he wrote a paper with Rosen on gravitational waves. In it they assumed plane symmetry and studied wave solutions (see figure 4.4); this particular symmetry simplified the equations enough to obtain an exact solution.

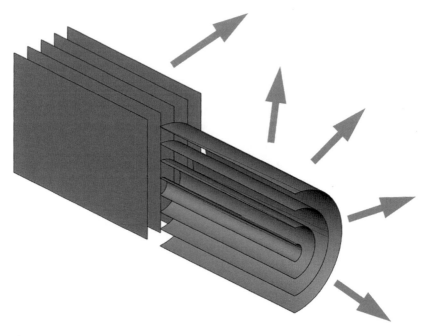

Figure 4.4
Plane and cylindrical waves propagating in the direction indicated by the arrows.
(Credit: Américo Hinojosa, University of Texas Río Grande Valley.)

In their paper, they noted that all the solutions had problems; although there were waves, certain quantities became infinite. So they concluded that these solutions did not represent true physical situations (in nature, nothing can be infinite) and therefore gravitational waves . . . did not exist! The paper was sent to the journal *Physical Review*, an American publication whose prestige had been growing. Eventually it would become the most prestigious physics journal after World War II, when the predominant center of physics research moved from Germany to the US. This increase in prestige happened under the watch of the legendary editor John Tate. He was the one who received the Einstein–Rosen paper and sent it for evaluation to the American physicist Howard Percy Robertson, professor at Princeton University. Robertson, who was on sabbatical at the California Institute of Technology in Pasadena, was one of the few experts on Einstein's theory in the US. He noted that although the math in the paper by Einstein and Rosen was correct, the physical interpretation was not. They thought they had solved the equations for plane waves, but in reality, they have done it for cylindrical waves. The quantities that became infinite were not physical quantities but mathematical artifacts of how the calculation was done. Robertson wrote a referee report, sent it to Tate, and the latter forwarded it to Einstein. Tate's letter did not reject the paper by Einstein and Rosen, it simply requested that they take into account the comments of the referee. The name of the referee was kept anonymous in the *Physical Review*. Today this is standard practice in all journals, but the German journals where Einstein had published until he came to the US did not follow it. Tate did not always send papers out for refereeing. The previous paper of Einstein with Podolsky and Rosen was not refereed. So it is possible that this was the first time Einstein had been faced with a referee report. Perhaps this explains why he reacted rather furiously, sending a pointed letter to Tate saying that he had sent the paper for publication and had not authorized him to show it to someone else. Tate replied that this was standard practice, but Einstein withdrew the paper and submitted it to a rather obscure journal in Philadelphia, called the *Journal of the Franklin Institute*, which promptly accepted it. Apparently he did not even read the referee report.

In the meantime, Robertson had returned to Princeton after his sabbatical and went to lunch with a new Polish assistant of Einstein (and future father of Polish physics in the postwar period), Leopold Infeld. Infeld mentioned to Robertson the results of Einstein and Rosen. Robertson pointed out the problems with the work—without revealing that he was the referee. Infeld ran back to Einstein's house and explained them to him. Einstein quickly understood. He was supposed to give a talk the following day titled "Do gravitational waves exist?" He gave a rather opaque presentation and concluded that the subject was so confusing that he did not know the answer.

He amended the paper in the *Journal of the Franklin Institute* before publication and the final version does not question the existence of gravitational waves. Although the original version has not been preserved, Rosen had moved to the Soviet Union and submitted a paper to a conference that was most likely similar to the original submission, since he had not heard of the objections of Robertson at the time of the conference. Rosen's published paper claims that gravitational waves do not exist.

Unbeknownst to Einstein and Rosen, the solution for cylindrical waves had been considered in a PhD thesis in 1923 by Austrian physicist Guido Beck (who, after escaping the Nazis in World War II, would be influential in the development of physics in Argentina and Brazil). Einstein himself admitted being "lazy" and not checking carefully the literature before conducting research.

4.6   The Low Tide of General Relativity

The period from 1925 to 1955 has been dubbed by the French physics historian Jean Eisenstaedt as "the low tide of general relativity." Essentially there were no experimental predictions that could be verified. Also, many relativists had followed Einstein's steps in trying to develop a unified theory of classical electromagnetism and gravity, something that proved futile when it became clear there were other interactions in nature and the description had to be quantum mechanical. In the US, universities stopped hiring relativists, considering that they were mathematicians rather than physicists. Relativists had to find jobs in teaching institutions. In the British Commonwealth, something similar happened, but relativists were able to find jobs in departments of mathematics. This had a rather positive influence in the field. New mathematical techniques were introduced that allowed researchers to understand the theory much better. Among other important developments, the meaning of the Schwarzschild solution was finally understood, giving rise to the modern concept of a black hole. It was also during these years that it was rigorously proven that gravitational waves carry energy away from systems, finishing once and for all the arguments of "traveling at the speed of thought." The generalization of the black-hole solution to the rotating case (crucial in astronomical applications, as all astronomical objects rotate) was found near the end of this period by New Zealand physicist Roy Kerr.

In the US, the "low tide" had an unpredictable and very important impact on the detection of gravitational waves. Given the lack of academic jobs, a group of relativists, including Josh Goldberg, Jeff Winicour, and Richard Isaacson, found jobs at the Wright–Patterson Air Force Base near Dayton, Ohio. It was rumored that the Air Force wanted relativity investigated to see whether anti-gravity could be produced, something obviously useful for the Air Force. No physicist seriously thought this was possible, but it would not be the only time that the US armed forces conducted dubious research. A 2009 film, "The Men Who Stare at Goats"—directed by Grant Heslov and starring George Clooney, Ewan McGregor, Jeff Bridges, and Kevin Spacey—was a satirical comedy with an ironic view on this type of research. It was based on a book by Jon Ronson about the exploration of "new-age" concepts by the US Army related to paranormal effects. The relativists employed at Wright–Patterson helped the Air Force avoid funding ideas without scientific support on the subject of gravity. And their research was first rate. The Mansfield Amendment in 1969 prohibited military funding of research that lacked a direct or apparent relationship to specific military function and led to the disbandment of the relativity group at Wright–Patterson.

Goldberg secured a position at Syracuse University, in New York State, where he retired as an emeritus professor until his death in 2020. Winicour went to the University of Pittsburgh, where he is a research professor at the time of this writing. The authors of this book spent time with these individuals during postdoctoral or doctoral studies and benefited from their knowledge and wisdom. Isaacson—who specialized in gravitational waves—took a different path in his career. He secured a job at the National Science Foundation, a scientific funding agency. Given the drop in military funding for research, the number of proposals going to the National Science Foundation increased significantly, and they needed more program officers to handle them. Isaacson's presence meant that gravitational physics always had a prominent place at the Foundation. He played a key role in funding early experiments on gravitational wave detection, and eventually the Laser Interferometer Gravitational-Wave Observatory (LIGO) project. At that time, program officers had great latitude in their decisions, which allowed Isaacson to proceed in spite of the initial skepticism that the project attracted—particularly from astronomers who considered the project too risky. Isaacson has stated publicly that today a project like LIGO would probably not be funded, given the more stringent bureaucratic rules in place. With a generous donation from 2017 Nobel Prize winners Rainer Weiss and Kip Thorne, the American Physical Society has recently created in 2018 the annual Richard A. Isaacson Award on Gravitational Waves, recognizing the key role he played in their detection.

The 1960s witnessed a renaissance of general relativity, particularly in the US. This was due to the discovery of quasars, astronomical objects that contain a source of energy so big that it can only be explained as a rotating black hole (we will return to this topic in section 4.7). This stimulated investigation in general relativity, and as a consequence, the debate on gravitational waves resurfaced. In spite of the mathematical advances that showed they were a real physical phenomenon that could transmit energy over distances, a significant number of theoretical physicists still doubted that black holes existed. The controversy was illustrated by several discussions in one of the first conferences on general relativity in Chapel Hill, North Carolina, organized by Bryce DeWitt and Cécile

DeWitt-Morette in 1957. The latter was one of the first women in relativity and had to overcome several obstacles to develop her career. The Chapel Hill conference was the second large international conference in the field. The first one was in Bern, Switzerland, celebrating the jubilee of special relativity in 1955. The intention was obviously to celebrate Einstein as an honored guest, but he passed away a few months before the conference took place. A series of meetings were initiated to take place every 3 years, which is still ongoing. The meetings, now organized by the International Society of General Relativity and Gravitation, are colloquially known as the "GR" meetings. At the 1957 Chapel Hill meeting, Richard Feynman participated in several discussions concerning gravitational waves. He became quite exasperated with the objection of theorists and in a memorable letter to his wife, he wrote "remind me not to come to any more gravity conferences."

Skepticism about gravitational waves continued a bit longer. At an international conference in the late 1960s, the American physicist Ted Newman (who passed away in 2021) chaired a session and took a poll asking whether a binary system emits gravitational waves. Half of the room said "yes" and the other half "no." Eventually the controversy was settled experimentally—quite appropriate for physics—by the observation of a binary pulsar, a story that deserves its own place in this book in the next section.

The relativity renaissance had its nerve center at Princeton University in the hands of John Archibald Wheeler, who had worked on the Manhattan Project and shifted his research interests from quantum physics to general relativity. He was responsible for popularizing the term "black hole." He also educated a generation of American relativists that populated the US universities and some in other countries, who became leaders of the field and are now nearing retirement. His students included Kip Thorne, one of the recipients of the Nobel Prize for the detection of gravitational waves.

4.7   Gravitational Waves Do Exist!

The first observational proof that gravitational waves exist happened in the 1970s, after much progress observing properties of stars using radio waves.

After the 1930s, radio astronomy became a new transformational tool in astronomy. Until then, astronomers had only used telescopes that detected visible light from stars. The start was serendipitous: Karl Jansky, an engineer at Bell Telephone Laboratories, was asked to investigate certain noise sources in radio antennas that were to be used in transoceanic communications. Noise can be defined for an instrument as any output registered in the instrument that differs from the expected signal.[2] He was surprised to discover that one of the noise sources came from the direction of the Milky Way in the sky. Jansky's discoveries motivated several astronomers to use radio antennas similar to his to explore the sky. Two discoveries from radio astronomy would be revolutionary: quasars and pulsars.

Toward the end of the 1950s, mysterious sources of radio waves were discovered in the initial surveys of the sky, being some of the first sources not to be clearly associated with visible phenomena. Eventually, it was possible to connect them with objects that looked like stars, but these objects were very far from the Earth and from our own galaxy, therefore the amount of energy being emitted was very large. The resemblance to stellar sources led astronomers to call them "quasi-stellar objects" or "quasars." Today it is known that these quasars are galaxies that are very far away, and their colossal energy is produced by disks of matter rotating at great speeds around black holes in their center.

At the end of the 1960s, the Northern Ireland astronomer Jocelyn Bell Burnell and the English radioastronomer and 1974 Nobel Prize winner Antony Hewish discovered another notable source of radio waves while looking for more quasars. The mysterious source emitted radio pulses every 1.33 seconds. A detailed analysis showed that the signals did not originate on the Earth. For this reason—with a clear sense of humor—the first signal was designated LGM-1 for "little green men." Together with similar discoveries, the sources began to be called by a combination of the words "pulsating" and "quasar": pulsar. After the discovery of other pulsars, it became clear that they were very compact stars mostly composed of neutrons.

A neutron star is essentially a stellar corpse, the remnant of a star that has exhausted its nuclear fuel and collapsed into a volume of a few kilometers in diameter and has close to 1.5 times the mass of the Sun. They emit radio beams that are not aligned with the star's rotation, so they behave like lighthouses. We see repeating pulses when the radio beam points toward the Earth (see figure 4.5). These are the pulses that radio telescopes detect from pulsars.

Figure 4.5
Schematic drawing of a pulsar, in which one sees the lines of magnetic field, the rotation axis of the star, and the beams of emitted radio waves at 45 degrees. (Credit: Wikimedia Commons, Mysid, CC-BY-SA-3.0.)

Joseph Taylor at Princeton University proposed to his PhD student Russell Hulse a thesis project looking at many pulsars to find a system formed by two stars: a binary pulsar. Their motion could be used to test the theory of general relativity. Finding whether a pulsar is in a binary system can be done by looking for small variations in the times of arrivals of the pulses, due to the motion of the pulsar orbiting around the other star. This could be done using the 300 m radio telescope at Arecibo, Puerto Rico (which, sadly, collapsed in 2020 after 53 years of many radio discoveries). The project was successful: they discovered the first binary pulsar in 1974. By 1979, measurements showed that the orbit was shrinking, which is due, according to Einstein's theory, to the gravitational waves taking away energy from the system. As they get closer, the stars move faster (just as ice skaters do by bringing their arms closer to their bodies), and the emission of gravitational waves increases. Since this effect is cumulative and can be measured for a long time, it is possible to do so with great precision. After more than 10 years of measurements of the shortening of the rotation period, the agreement with the theory of general relativity was 1 percent. The British physicist and mathematician and 2020 Nobel laureate Sir Roger Penrose has said

> Indeed, if we now take the system as a whole and compare it with the behavior that is computed from Einstein's theory as a whole—from the Newtonian aspects of the orbits, through the corrections to these orbits from standard general relativity effects, right up to the effects on the orbits due to loss of energy in gravitational radiation—we find that the theory is confirmed overall to an error of no more than about 10^{-14}. This makes Einstein's general relativity, in this particular sense, the most accurately tested theory known to science![3]

The discovery led to a Nobel Prize for Hulse and Taylor in 1993. Russell Hulse was a PhD student when he did this work; afterward he moved into plasma physics at the Princeton University Plasma Physics Laboratory to be near his wife. The Swedish Academy rightly recognized him with the prize: he and his advisor had discovered the system together. In the past, in similar cases, like that of Jocelyn Bell Burnell, who in her PhD work discovered pulsars, the prize went to the supervisor and a colleague, but not to the student.

The Hulse–Taylor binary pulsar system provided strong and precise evidence about the existence of gravitational waves, even without detecting them directly. This evidence finally ended the controversy among theorists about the existence and production of gravitational waves.

5

The Life and Death of Stars

As Einstein wrote in his 1916 article, gravitational waves are very weak. According to his theory, they are produced by accelerated nonspherical mass distributions. Their effect will be nonnegligible when the accelerations, masses, and velocities are very large. This does not happen in terrestrial experiments, but it can happen in the skies, with stars exploding or moving very fast, like the neutron stars in the Hulse–Taylor binary pulsar, where the energy lost to gravitational waves can be measured (over many years).

Before attempting a description of the various astrophysical sources of gravitational waves, we will review the natural history of stars, as this history plays a crucial role in these sources.

5.1 The Birth and Life of Stars

The universe started in a very hot and dense state consisting of the elementary particles known as quarks and electrons. As the universe cooled down, the quarks clumped together to form protons and neutrons, which combined with the electrons to form the lightest atoms: fundamentally hydrogen and also helium in a smaller quantity. These atoms then attract each other and form "molecular clouds," which are still forming in most galaxies, including ours. In the hydrogen cloud, gravity starts its work, and the interior becomes very dense; atoms are bouncing off one another and moving more quickly: the temperature is rising (if an object is hotter than another one, it is because its atoms are moving more quickly). When the center of the forming star becomes sufficiently dense, temperatures inside it reach millions of degrees, enough to start thermonuclear fusion. In a fusion process, pairs of hydrogen atoms merge to form heavier helium atoms, liberating in the process an extra amount of energy, predicted by Einstein's famous formula $E = mc^2$.

The atoms at the center move faster than the atoms farther away, so they move away from the center, creating an outward pressure, and the tremendous heat generated at the center is then transported to the star's surface. Equilibrium is then reached between the expanding heat pressure of the star's core and the attractive force of gravity. This equilibrium lasts as long as the star's source of nuclear fuel, hydrogen, lasts. This process was first proposed by Sir Arthur Eddington (the same Eddington who proved that Einstein's theory was right with the 1919 eclipse observations), who published in 1920 a famous paper in astronomy titled "The internal constitution of the stars," even though at the time thermonuclear fusion had not been discovered.

When the star runs out of fuel, it suffers catastrophic changes that will affect its nature. Depending on the mass, the process can last from a few hundreds of millions to a few billion years, so there's no need to worry: this won't happen for our Sun for another 4–5 billion years. You might think that once the gravitational force cannot be stopped, the star collapses immediately into an infinitely compact object. This is not quite true, which is the origin for the rich variety in the lives (and deaths) of stars. Depending on the star's mass, various possibilities arise.

The helium atoms at the core of the stars will begin fusing into heavier carbon atoms. For our Sun and other stars that are up to about eight times as massive as the Sun, there is a stage when some hydrogen from the enveloping layers is gravitationally attracted to the core, and these atoms begin thermonuclear fusion again. This "revival" makes the star expand, which increases several times in volume and also cools it down. The cooling makes the starlight more reddish, forming a *red giant*. Eventually red giants run out of fuel to keep the fusion process alive, and gravity keeps contracting the star.

If the star is more massive, the fusion in the core continues beyond carbon, until it forms iron atoms, the most stable form of nuclear matter. Now gravity wins the race again, and there is a much more violent and explosive collapse: a supernova explosion.

Inevitably, the final result of the process is a dead star. Depending on the mass of the star, the remains can be of three different types: white dwarfs, neutron stars, or black holes. We will explore each of these types in some detail.

5.2   Small Masses: White Dwarfs

White dwarfs were found for the first time in binary systems. Their motion around each other as measured observationally allowed astronomers to infer their properties, such as their masses. The most famous example, shown in figure 5.1, is Sirius, discovered to be a binary system in 1844. Sirius A is the most brilliant star at night; its companion, Sirius B, was found to have a bit less mass than the Sun but is much, much less luminous—a dying or dead star. However, spectral (color of light) measurements in 1914 showed that Sirius B is white and hot, not cool; the small luminosity then implies it is a very small star: a white dwarf. Its density had to be so high that it generated much incredulity among many, including Eddington:

> The message of the Companion of Sirius when it was decoded ran: "I am composed of material 3,000 times denser than anything you have ever come across; a ton of my material would be a little nugget that you could put in a match-box." What reply can one make to such a message? The reply which most of us made in 1914 was— "Shut up. Don't talk nonsense."[1]

Figure 5.1

Artist's conception of the binary star system Sirius: Sirius A is the most brilliant star at night and its companion, Sirius B, is a white dwarf (shown on the right in the picture, much smaller than Sirius A). Sirius is approximately 8 light-years from the Earth. (Credit: NASA, STScI.)

Eddington himself came up with the explanation. During their life, all stars are made of atoms, which means they are mostly empty, since atoms themselves are also mostly empty: if an atomic nucleus was as big as a tennis ball, the electron orbits would be the length of a football field. But the electrons liberated in the star now end up preventing a bigger collapse, due to an exotic quantum effect called "electron degeneracy pressure," a repulsion among electrons much stronger than Coulomb's electrostatic repulsion. The application of this electron pressure stopping the collapse of a white dwarf was postulated by Eddington in 1924, even though quantum mechanics and its interpretation were still in their infancy. He went further, suggesting another observational test of Einstein's theory: measuring differences in the gravitational redshift due to the rotation of the small star, a test that was done successfully only a year later. Eddington proved Einstein right twice!

However, this explanation only works up to some limiting size for the star, a limit that is known as the *Chandrasekhar limit*. Chandrasekhar's observation was that if the mass of the star is greater than 1.4 solar masses, the electron pressure would not be enough to stop gravity from merging the heavier protons and neutrons in the star, and the star will collapse further, becoming so compact that even light cannot escape from it. Subrahmanyan Chandrasekhar was an Indian astrophysicist who was born in Lahore, Punjab (modern Pakistan), and studied at the University of Madras in Chennai before moving to the United Kingdom to complete his PhD. He received the Nobel Prize in Physics in 1983. Chandrasekhar derived this limiting value based on purely theoretical considerations. He faced discrimination due to his origin throughout his life. In England in the 1930s, he worked under the supervision of Eddington, who disliked and opposed his work. Chandrasekhar talked about these results in 1935 at a meeting of the Royal Astronomical Society, after which Eddington stated in the talk that followed in the same meeting "I think there should be a law of Nature to prevent a star from behaving in this absurd way." Eventually Chandrasekhar left Europe and moved to the University of Chicago for the rest of his career. During World War II, he was invited to take part in the Manhattan Project, but delays in his security clearance prevented him from joining it.

In spite of exhausting their source of energy, white dwarfs stars keep on shining due to the residual temperature they have. But ultimately, they will start slowly dimming until they become completely dark. Astronomers think that no white dwarf in the universe has yet become completely obscure: the universe is not old enough. Current galactic models indicate that there are about 10 billion white dwarfs in our Milky Way. And it is also known that more than 90 percent of its stars (including our Sun) will end their lives as white dwarfs. But as Chandrasekhar noticed, there should be even more compact objects in the universe.

5.3   Medium Masses: Neutron Stars

Hans Bethe—the German–American physicist who won the Nobel Prize in Physics in 1967—and other scientists were able to explain the source of the energy of stars as being thermonuclear fusion. The neutron—which together with the proton, are the fundamental constituents of the atomic nucleus—was discovered in 1932. Shortly thereafter, the Swiss astronomer Fritz Zwicky and the Soviet physicist Lev Landau (mentioned in chapter 4) considered the possibility that stars could exist composed entirely of neutrons. But 20 years would still pass before observational astrophysicists identified these stars that have incredible densities associated with cataclysmic processes in the universe.

Stars with more than eight solar masses end their life cycle in an explosive collapse of their core, known as a supernova (see figure 5.2). At their center, all that would be left is a neutron star, a black hole, or nothing.

Even after ejecting a large fraction of its mass during the explosion, the mass of the remnant star ends up exceeding the Chandrasekhar limit. The pressure of the free electrons is not able to counter gravity, and the star collapses even further. How much more? The free electrons collide with the protons, and in a process known as electron capture, they create neutrons. In this reaction, lighter particles are emitted, known as neutrinos. Neutrinos can be detected with special observatories (see figure 5.3). In 1987, a supernova explosion was observed with optical telescopes, and an associated emission of neutrinos was also detected.

Figure 5.2
The supernova SN 1994D (the brilliant flash at the lower left). Photo of the NGC 4526 galaxy, where the star that originated the supernova exploded, taken by the Hubble Space Telescope (Credit: NASA.)

Eventually, as the star collapses after the explosion, a core composed mostly of neutrons is created, and their gravitational attraction continues to collapse the core. However, there is a limit to the star collapse, again due to quantum mechanics. Neutrons are also particles that, like electrons, have a degeneracy pressure that can balance gravity, as long as the mass of the star is less than about 2.5 times the mass of the Sun. The resulting compact object has a density of about 100 trillion times that of water. This star has a

Figure 5.3
The IceCube neutrino telescope in Antarctica. Buried under the ice are arrays of vertical sensors longer than a kilometer. (Credit: Felipe Pedreros, IceCube/NSF.)

diameter of around 6 miles: the size of a small city. It is a "neutron star" (although there are heavy atoms in the star, too, not just neutrons). The first observations of neutron stars were made at the end of the 1960s by Jocelyn Bell Burnell and her advisor Anthony Hewish, and they were used to prove for the first time the existence of gravitational waves with the Hulse–Taylor binary system.

5.4   Extreme Masses: Black Holes

The Schwarzschild solution was the first-known exact solution of the Einstein equations. It represents the gravitational field of a spherical object and predicts that the density of energy in its center becomes infinitely large: a singularity. In an article written in 1939, Einstein stated that "the Schwarzschild singularity does not appear for the reason that matter cannot be concentrated arbitrarily. And this is due to the fact that otherwise the particles that constitue the matter would reach the velocity of light." In spite of this, that same year, the famous physicist Robert Oppenheimer (leader of the Manhattan Project, which created the atomic bomb) and his student, Hartland Snyder, published a new result. They showed that the solution could be used to model the collapse of a star when its source of energy is extinguished, but the meaning of their work was not completely understood until later.

Together with other physicists, Oppenheimer realized that neutron stars could not exist with a mass larger than a value that they originally estimated as slightly lower than the mass of the Sun. It was later computed to be about twice that mass. A star that exhausts its nuclear fuel and ends up with a mass larger than that limit does not have any internal pressure that can counteract gravity. Above that mass limit, gravity is unstoppable, and matter is completely squeezed into an infinitesimally small volume. This limit, similar to the one deduced by Chandrasekhar, is called today the Tolman–Oppenheimer–Volkoff limit. With a more rigorous knowledge of nuclear physics, the pressure exerted by neutrons can now be more precisely studied, and it is estimated that the value of the limiting mass is somewhere in between two and three solar masses. It was then clear that certain stars could collapse and become black holes, as described by the Schwarzschild solution. It was not a mathematical result without connection to real astrophysics anymore. The first searches for black holes started in the 1960s.

In 1964, a strange source of X-rays was found in the constellation of Cygnus, and it was named Cygnus X-1 (see figure 5.4). The mass of this object is estimated to be 15 solar masses. Its size is such that the stellar object is confined to a radius smaller than 200 miles. This object is denser than a white dwarf or a neutron star, and it is the first evidence of the existence of a black hole! But how could scientists obtain all this information if the system does not emit light?

Figure 5.4
The black hole in Cygnus X-1 in this artists' conception is to the left, at the center of the disk. The black hole rotates around a star, from which it absorbs matter. (Credit: NASA/CXC.)

Cygnus X-1 is part of what is known as an X-ray binary system. It is about 6,000 light-years from the Sun and includes a star known as a blue giant (due to the high temperature of its surface, this star color is bluer than a much smaller one would be). The black hole orbits around it at a distance equivalent to half the distance between Mercury and the Sun. The stellar wind of the star (i.e., the flux of particles from the stars' atmosphere) is captured by the black hole, forming a disk—known as accretion disk—in which material rotates at very high velocities. Due to the extreme temperatures generated, the accretion disk emits X-rays, as shown in the figure. These X-rays allow astronomers to observe the system, although they cannot see the black hole directly.

In 1974, the famous English physicist Stephen Hawking placed a bet against Kip Thorne that Cygnus X-1 was not a black hole; in

1990, Hawking conceded it. After Cygnus X-1, several similar systems were found. There are more than 20 stellar black holes now known in our galaxy, all discovered through the studies of their associated X-ray emissions. These black holes are members of a type of black hole called "stellar black holes." Their masses are more or less similar to that of known stars. Their progenitors were most likely stars with masses larger than a few solar masses.

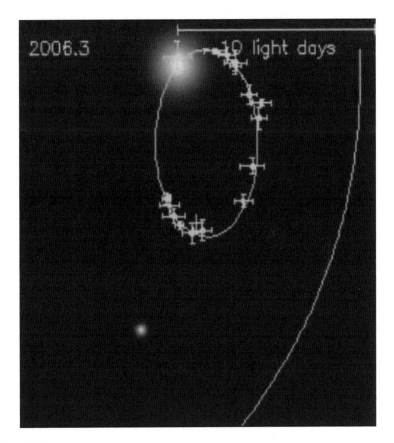

Figure 5.5
Measurements by Andrea Ghez's group of the orbit of star S2 around the black hole at Sagittarius A*. The black hole is not visible; it is located at a focus of the ellipse. The bright spot is the star at one of its positions, the other positions (indicated by crosses) are for the years between 1992 and 2006. (Credit: Wikimedia. European Southern Observatory.)

But there are other types of black holes. Supermassive black holes, huge behemoths with masses several million times the mass of the Sun have been discovered at the center of almost all known galaxies. How they form is still a mystery, but there are good reasons to think that the process is very different from the formation of stellar black holes.

The Nobel Prize in Physics 2020 was a recognition and celebration of black holes. An enigmatic and almost unfathomable astrophysical relic has finally come of age. The prize was awarded half to Roger Penrose and half to both Reinhard Genzel and Andrea Ghez. Penrose received it for showing in 1965 that the general theory of relativity leads necessarily and unavoidably to the formation of black holes. Reinhard Genzel and Andrea Ghez each lead a group of astronomers that, since the early 1990s, has focused their observations on a region called Sagittarius A* at the center of our Milky Way. They both studied independently the orbits of the brightest stars around it with great precision. The measurements of these two groups agree, both finding an extremely heavy, invisible object that has a mass of about 4 million solar masses packed in a volume no bigger than our solar system. In particular, one star, called S2 or S0-2, with a mass estimated to be around 14 solar masses, was observed to complete an orbit around this object in about 15 years. Its motion can reach speeds higher than 5,000 km/sec (1/60 of the speed of light!), making it the fastest celestial object ever recorded. It follows with incredible precision the orbit predicted by the Schwarzschild solution of Einstein's equations of general relativity.

And this is how stars die, littering with their corpses—white dwarfs, neutron stars, and black holes—the immensity of the cosmos.

6

Astrophysical Sources of Gravitational Radiation

Just as electromagnetic waves have many different frequencies, from X-rays to the visible spectrum to microwaves, gravitational waves also have a wide range of frequencies and wavelengths. In general, the larger the size of the source, the larger the wavelength and the lower the frequency. But there are also different kinds of waves: here we describe some of their astrophysical sources.

6.1   Stellar Binary Systems

Many observed stars are in binary systems, composed of two stars orbiting around each other. According to Einstein's theory, any binary system will generate gravitational waves. Since they carry energy away from the system, the stars will get closer and eventually merge. The stars slowly dance a serene waltz around each other; the frequency of the gravitational radiation emitted will start low (proportional to the time the stars take to orbit), but it will increase. Near the end of the dance, the waltz become more like a fierce tango with the frequency increasing, very shortly before the dancing stars embrace each other and form a new system. The final system will be rotating, and it will still produce gravitational waves unless it is a black hole.

The type of signal generated in the last instants of a binary system's life is a periodic humming that, slowly at first and then more quickly, increases its frequency and amplitude. At the instant the stars merge, the amplitude of the wave reaches its peak and then decays like the ring-down of a bell after being struck (see figure 6.1). The smaller the masses, the higher the final frequency and smaller the final amplitude will be (think of a small bell: it has a high pitch).

If the binary system is made of compact objects like neutron stars or stellar black holes with diameters of ten to hundreds of kilometers, the final frequency will be in the band of terrestrial gravitational wave detectors such as LIGO and Virgo. If the masses are

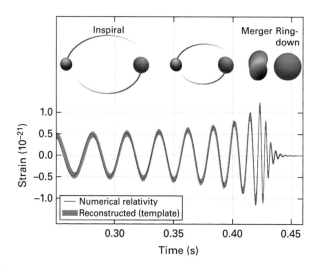

Figure 6.1

The gravitational waveform for a binary system that coalesces. As the black holes get closer, the signal frequency increases (the waves get more closely packed together) and the amplitude increases as well. After the merger, the amplitude decays quickly. The gray line represents the theoretical template that best fits the data, and the dotted line is the result of a numerical solution of Einstein's equations for the masses observed. (Credit: B. Abott et al., "Observation of Gravitational Waves from a Binary Black Hole Merger." *Physical Review Letters* 116, 061102 (2016).)

larger, like massive black holes at center of galaxies or white dwarfs that have the size of planets or larger, the frequencies will be detectable by larger space or galactic-size detectors. We describe such detectors in detail in chapter 12.

Since we know what the signal looks like (especially before the merger), we can use a technique called "matched filtering" to look for the astrophysical signal in the output of a detector. If the detector's noise is random in nature, this is known to be the optimal method, and it is routinely used with radars and sonars, where a signal is generated (and therefore known), and the reflected signal is looked for in the receiver.

How well can we predict the gravitational waves emitted by the system? Einstein's theory predicts them, but its equations cannot be solved with absolute precision. However, approximate techniques yield very good results, at least before the final crash. These methods are known as post-Newtonian, because they start with Newtonian physics describing the motion and study the small differences between such solutions and those from Einstein's equations. The waveform near the merger can be supplemented with numerical methods like the ones we discuss in chapter 7. Wave patterns or profiles representing many different possible solutions can be created with pre-established values of the masses and other physical parameters (initial velocities and rotation rate of the object, their positions in the sky, etc.) that we want to determine. A large number of these patterns, called "templates," are arranged in a "bank," which is compared one by one with the data from the detector at the particular time when the signal is suspected to have occurred. The technique, which measures how good a given template matches the signal detected, is known as "matched filtering."

The data obtained from the detector—even that containing a signal—always have instrumental noise, as we discuss in chapter 9. If there is no signal, the matched filter will produce low and similar numbers for all the templates in the bank. If there is a signal of significant amplitude, the filter will give a maximum value for the template of the bank that has the closest shape to the signal. The value will be considerably larger than the one obtained when no signal is detected. This indicates that one has a candidate for detection.

Many mergers of binary systems have been detected, and we will discuss them in chapter 10.

6.2   Rotating Stars

Any rotating nonspherical star—for example, a neutron star with a small mountain in its crust—will be a source of gravitational waves. All stars rotate, and they are not perfectly spherical. Neutron stars nearly are, with a radius of about 6 miles but protuberances that are smaller than half an inch! The gravitational wave will be sinusoidal, with the frequency depending on the star's rotation speed; the amplitude will be determined by the nonsphericity of the star and its distance from the detector.

Most stars rotate. For instance, our Sun has a rotation period of about 27 days. These low frequencies, coupled with small amplitudes, lead to signals that are very difficult to detect. But some stars, like pulsars, rotate much more quickly. They have periods from less than 10 seconds to just 1 or a few milliseconds (those are called "millisecond pulsars"). The gravitational waves emitted by the latter are in the sensitive band of the LIGO and Virgo detectors.

How many pulsars are there, and how far away are they? More than 2,000 pulsars have been found in our galaxy, and all of them are hundreds of light-years away. But these have been detected, because luck has it that our Earth is in their beams' paths. Pulsar beams are quite narrow, and there are likely many more of them pointing randomly in directions that would not reach us. Astronomers think there are hundreds of millions of neutron stars in our galaxy, and some of them may be closer to us.

The gravitational radiation will be of a quite stable frequency that is known for every pulsar observed with radio telescopes. Their localization in the sky is also known. Having a known localization and frequency allows us to search for this signal in the data of the detectors. The advantage is that, since the signal is continuous, it can be studied for a long period of time. However, no continuous signals have been detected so far in the data of LIGO and Virgo.

To search for the pulsars not pointing toward Earth, we can use the matched filtering technique for all possible frequencies and positions in the sky of the source. That makes it, however, computationally very costly. This led to the creation of a project called "Einstein@home," which distributes a screen saver to the public to install on their computers. When the computer is not being used for other purposes, the program carries out portions of the calculation necessary to find neutron stars without knowing their frequency or position in the sky. Interested readers can download it at einstein-athome.org. You can be part of the next discovery!

6.3   Supernova Explosions

In a supernova explosion, great quantities of the star mass are violently shaken, and a significant amount of matter is also ejected. This acceleration of matter generates a burst of gravitational waves, although of less intensity than a collision of compact objects of similar mass. The waveforms produced during the explosion are not known with precision. Although some progress has been made in the numerical modeling of these bursts, we cannot take advantage of the matched filtering method. These searches must be done looking for an excess of energy in the detector. Looking for coincident signals in more than one gravitational wave detector is essential, given the lack of models. But we can also use information provided by astronomers: when they observe supernova explosions in the sky through more traditional telescopes, we can look for excess energy in the gravitational wave detectors. A supernova explosion produces gamma rays, X-rays, visible light, and even neutrinos, although these and gravitational waves are expected to be detected only if the explosion happens in our galaxy. We hope this happens soon!

Of course, there may be other sources of transient, unmodeled gravitational waves; the methods used could find a transitory signal that appears in LIGO/Virgo even if there is no other kind of astronomical observation. Every time new instruments have been developed to observe the universe, surprises appear and lead to the discovery of new phenomena. It happened with radio astronomy; gamma ray detectors; and the optical telescope itself, when Galileo pointed it up to the sky for the first time more than 400 years ago and discovered that Jupiter had moons. It is possible that the same will happen with gravitational wave detectors. Time will tell.

6.4   Stochastic Background of Gravitational Waves

The current model for the universe predicts that gravitational waves come from all directions in the sky, mix with one another, and appear as a noise of cosmic origin. They are not transient signals, they are present all the time, but they do not have a definite pattern or structure. They are known as "stochastic" (another word for "random") signals.

A very well-known example of a signal of this type in the electromagnetic spectrum is the cosmic microwave background. Waves in the microwave band have wavelengths between 1 mm and 30 cm; they include waves from radar, GPS, Bluetooth, and the ones generated by your microwave oven. The latter have 12 cm wavelength; the cosmic microwave background waves that originated shortly after the Big Bang are in the same band and have their largest amplitude at 1 mm wavelength. They were discovered during a continuation of the investigations of Jansky described in section 4.7.

Scientists at Bell Labs continued studying radio astronomical sources in the following years, and in 1964, two of their radio astronomers, Arno Penzias and Robert Wilson, detected a radio signal in the microwave band that they could not explain. At first they attempted pedestrian explanations, such as the signal was from bat and dove droppings on the radio receiver. Eventually it was understood that they were actually observing signals that came from the cosmos. Such signals were produced as soon as the universe cooled enough for light to be able to travel through it without being completely absorbed, about 300,000 years after the Big Bang. The earlier universe was a thick soup, and light could not get through it. Penzias and Wilson were awarded the Nobel Prize for their discovery in 1978.

A similar cosmological stochastic background of gravitational waves is expected to exist as well. However, although the microwave background radiation provides information about what the universe was like some 300,000 years after the Big Bang, the gravitational stochastic background originated with the Big Bang itself. If it could be detected, it would provide information about the origin of the universe that we would not be able to obtain by any other means.

The search for this signal is done using one or more detectors simultaneously. If such an astrophysical signal existed, it would be the same in all detectors, and its strength would be amplified. In particular, the strength of the signal above the noise (the so-called "signal-to-noise ratio") would be considerably enhanced. The noise of the detectors, since it should be of a different nature for each of them, would not be strengthened by the addition. And with the signal being a continuous one, its prolonged observation would amplify the probability of finding it. Contrary to the continuous signals from pulsars, its detection is computationally simpler, since it comes from all directions in the sky with equal probability, so we do not need to figure out where it comes from.

The predicted background exists at all frequencies, so in principle, it is in the band of multiple kinds of gravitational wave detectors. Unfortunately, models also predict an amplitude that is many orders of magnitude smaller than the noise of the detectors. This kind of background has been searched for in LIGO/Virgo, and predictably, nothing has been found. However, there is another way to look for this signal: those early gravitational waves should have slightly modified the properties of the cosmological microwave background. Measurements of this signal have increased in precision, and new sensitive detectors in the South Pole may be able to "see" this imprint of the early universe.

In addition to the Big Bang, signals from many sources throughout the universe that are too far away to resolve individually would also add up to a stochastic background. For instance, collisions of very distant black holes and neutron stars will be many in number but low in amplitude, and they would not be distinguished from one another. This is known as the astrophysical (as opposed to cosmological) background of gravitational waves. This kind of background is expected to be observed in several kind of detectors, as we will describe in chapter 12. The observation of these signals is just a matter of time!

7
Numerical Relativity

As we have discussed, knowing the signal in advance gives a significant advantage in gravitational wave detection. Portions of the signal can be adequately modeled using post-Newtonian approximation techniques. However, having the full signal would be even more advantageous. This requires solving the Einstein equations of general relativity.

The Einstein equations are too complex to be solved using pencil and paper except for very simple situations. In particular, if we wish to study the collision of two black holes that orbit around each other emitting gravitational waves and getting closer until they collide, we are dealing with a complex, time-dependent problem. Worse, it is a problem with two widely different time scales: the quick motion around the orbit and the very slow shrinkage of its radius due to the gravitational wave emission. The Einstein equations for a problem like this have thousands of terms. Each equation would occupy several pages of packed notation. The only way to attack the problem systematically is to use computers. Hence many scientists expected that "numerical relativity"—the name given to this area of the discipline—would play an important role in the detection of gravitational waves. Numerical relativity is quite useful for understanding the limits of the approximations like the post-Newtonian one and to calibrate other semi-empirical approximations that are used to determine the parameters of the systems detected.

7.1   First Attempts

The first attempts to solve the Einstein equations with computers go back to the 1970s with the work of Bryce DeWitt at the University of Texas at Austin. Although he specialized in quantum issues, which are not directly related to solving the Einstein equations, he assigned a group of PhD students to study the problem: Larry Smarr, Kenneth Eppley, and Andrzej Čădež. This was quite a challenge: the best supercomputers at the time had less computer power and memory than a smartwatch of today. To simplify things, the group decided to attack the problem starting with the study of a head-on collision of black holes. It is very unlikely that such a precise alignment will happen in nature, but it simplifies the physics quite a bit. Such a collision happens relatively fast and requires considerably less memory and computing power than a long inspiraling collision that would more faithfully represent what actually happens in nature.

The students succeeded in making a simulation, but they could not verify how accurate the results were. To do such verifications typically requires two simulations, a coarser and a finer meshed one, to check that the results are consistent with each other. If they are consistent, then they are trustworthy. Unfortunately, the coarse simulation by itself used all the computing power available. Nevertheless, the students concluded that approximately 2 percent of the mass of the black holes is converted into energy radiated in the form of gravitational waves. It was a good beginning. Everyone expected that computing power would increase with time, and consequently, a realistic collision was going to be achievable in a few years.

7.2   The Grand Challenge

Larry Smarr went on to direct the National Center for Supercomputing Applications at the University of Illinois at Urbana–Champaign. He established a large group working on numerical relativity. Several other groups were established, inside and outside the US. These groups generated a lot of activity, but problems developed. When realistic simulations were attempted with the black holes orbiting around each other, the computer codes crashed without completing even a fraction of an orbit. The head-on collisions that had been studied did not crash the computer codes, because the collisions happened very fast and therefore did not provide any indication of what the problem could be in the more realistic scenarios. No one knew what was wrong.

It was the 1990s—20 years after the first simulations performed at Austin—and progress was sluggish in spite of the great increase in computing power. This was a drawback for the community, particularly in view of the importance attributed to having accurate simulations of gravitational wave emissions. In an effort to overcome the impasse, the National Science Foundation (NSF) created an alliance of nine numerical relativity groups called the "Grand Challenge Alliance" and awarded it a large grant in excess of a million dollars per year. This sum was considerable for a project that did not involve equipment (the supercomputing facilities utilized were part of national centers that were funded independently). Richard Matzner at the University of Texas at Austin led the Alliance, and Kip Thorne placed a bet against him that LIGO would detect the waves before the Alliance could simulate them.

Several strategies were tried to improve the simulations. One of them was to study the collisions from the point of view of an observer that rotates with the black holes. This eliminates the rapid rotational motion, allowing the simulations to be focused on the convergence of the two black holes, which is a much slower process. Another important new scheme was the treatment of the singularities appearing in the calculations. A black hole contains in its interior a "singularity," a place where all the matter is concentrated (a place where its density is infinite). Computers cannot deal with infinite quantities. But since the black hole interior cannot communicate with the exterior (that is the whole concept of a black hole), what happens inside is irrelevant to the production of gravitational waves. The new technique simply consisted of omitting from the calculations the interior of the black holes, a procedure known as "excision." But a crucial aspect of this technique was to check that the excised region really was inside the black holes. This verification is technically complicated, since it is not easy to describe the evolution of the simulation from the point of view of an observer co-rotating with the hole. There were several other very sophisticated developments that are too technical to discuss here. But nothing seemed to work . . . and the years kept going by.

7.3   The 2005 Surprise

Everything suddenly changed in 2005. A postdoctoral researcher at Caltech, South African physicist Frans Pretorius, who had been a student of Canadian physicist Matt Choptuik—known for his refined simulations—presented a simulation in which two black holes orbited without problems through the computational grid. They did so until they coalesced, emitting gravitational waves. At the beginning it was not clear why Pretorius's simulations were working and those of other groups were not. Pretorius's approach combined several novel ingredients and techniques that were not used by the other groups. This made the reproduction of his results difficult.

While the community was trying to understand this, a few months later, two events took place. The Argentinian physicist Pedro Marronetti—at that time at Florida Atlantic University, now working at NSF in the same position that Richard Isaacson (mentioned in section 4.6) had—was visiting the group at the University of Texas at Brownsville.[1] This was the numerical relativity group led by Italian physicist Manuela Campanelli, Argentinian physicist Carlos Lousto, and US postdoctoral researcher Josef Zlochower (this group is now at Rochester Institute of Technology). In discussions with Marronetti, the group tried various modifications to the code, and suddenly the simulations started producing results similar to those of Pretorius. Without knowing this, and essentially at the same time, a similar process took place at the group led by Joan Centrella at the Goddard Center of NASA. Both groups were quite surprised when at a conference in December, they discovered that both had achieved similar breakthroughs. In a short time, all numerical relativity groups had modified their codes, and simulations were running successfully. The binary black hole problem had been solved, after almost 30 years of stagnation. Kip Thorne conceded the bet to Richard Matzner.

The reader should not get the impression that the problem was solved fortuitously. Many developments by many groups had taken place over the years that were crucial for achieving successful simulations. It is just that in 2005, like the pieces of a giant puzzle, they all came together.

7.4 Expectations and Reality

The results from numerical relativity showed that the waveforms resulting from the collision of black holes are relatively simple. In particular, the initial moments of the collision are well described by the post-Newtonian approximation. The final moments are also well described by the "close limit" approximation, in which the two black holes are treated as a single distorted black hole. This approximation was developed by US physicist Richard Price and one of the authors of this book, Jorge Pullin, at the time at the University of Utah. It in turn had been further extended by John Baker (now at NASA-Goddard), Manuela Campanelli, and Carlos Lousto in the "Lazarus project" (so called because it revived numerical codes that had crashed). It had long been speculated that the collision of black holes in the regime halfway after the post-Newtonian spiraling in and before the resulting larger black hole was ringing down (modeled using the close limit approximation) could have been the source of some unexpected and surprising waveforms. Furthermore,

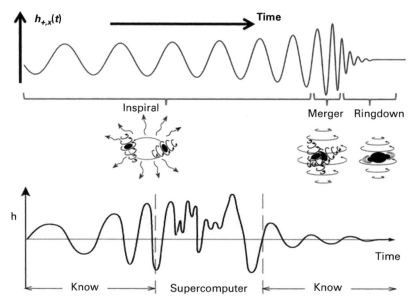

Figure 7.1
A numerical relativity waveform (top) compared to a waveform drawn by hand by Kip Thorne before reliable numerical simulations were available (bottom). (Credit: M. Favata/SXS/K. Thorne.)

it was also conjectured that those waveforms could not be described by any approximate technique. This skeptical prediction is illustrated in figure 7.1. But such expectations were not borne out. The waveforms can be almost entirely obtained by approximate techniques for most cases. This is fortunate, because the approximate techniques are computationally much less demanding than full numerical simulations. A full simulation can take up to a month of supercomputer time. In the end, a key role was played by an approximation developed by Alessandra Buonanno, currently at the Albert Einstein Institute in Golm, Germany, and Thibault Damour of the Institut des Hautes Études Scientifiques in Paris. This approximation was used by various physicists to develop semi-empirical formulas that could be calibrated using just a few numerical simulations and then used very efficiently to generate banks of templates of gravitational waves. These were the templates that were used to characterize the various parameters of the gravitational waves that were detected.

Plate 1

A view of M87 black hole in polarized light. A similar image appeared in the cover of most newspapers in 2019. (Credit: Event Horizon Telescope, https://www.eso.org/public/images/eso2105a/.)

Plate 2
Left: The Milky Way seen from Tolar Grande, in the foothills of the Cerro Macón, Salta, Argentina, where the astronomical park of the same name is located (photo: Samanta Fuentes). Right: The galaxy UGC12158, which is believed to be very similar to our galaxy, observed from outer space. (Credit: Hubble Space Telescope ESA/ Hubble & NASA.)

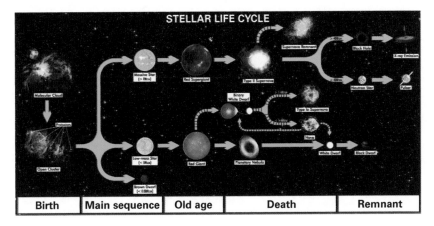

STELLAR LIFE CYCLE

| Birth | Main sequence | Old age | Death | Remnant |

Plate 3

Simplified diagram representing the paths of the life and death of stars, depending on the mass of the star. (Credit: NASA and schoolsobservatory.org.)

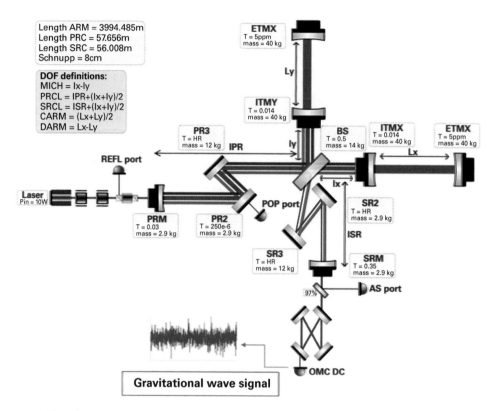

Plate 4

Schematic diagram of the beams and mirrors used in Advanced LIGO. The mass and designed power transmission T are noted for each mirror. HR stands for high reflector. PR stands for power recycling. PRC is the PR cavity. PRM, PR2, PR3 are three mirrors involved in the PRC. SR stands for signal recycling. SRC is the SR cavity; SRM, SR2, and SR3 are the mirrors involved in the SRC. AS, REFL, and POP ports are the antisymmetric, reflected and pick-off exit path for light detected on photocells, respectively. OMC is the output mode cleaner cavity; the photodiode with light transmitted by the OMC is calibrated in terms of strain to produce a gravitational wave data stream. (Credit: Anamaria Effler/LIGO.)

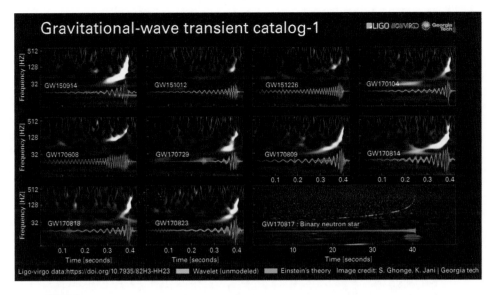

Plate 5

Time-frequency and waveform diagrams of the 11 gravitational waves detected in the first and second observing runs, 2015–2017, included in the first catalog GWTC-1. The first 10 are from collisions of black holes (all shown for a half-second). The last one, detected on August 17, 2017, is a collision of neutron stars, shown for tens of seconds. (Credit: S. Ghonge and K. Jani/Georgia Tech.)

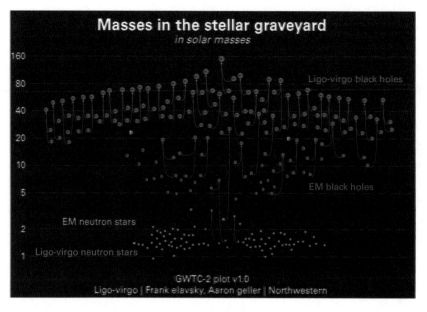

Plate 6

The black holes (blue) and neutron stars (orange) detected by LIGO up to the end of the O3 scientific run (March 2020). The figure shows all the known (through electromagnetic observations) black holes (in purple) but only a fraction of the known neutron stars (in yellow). The arrows indicate the final black hole created after the collision. (Credit: LIGO–Virgo/Frank Elavsky, Aaron Geller/Northwestern.)

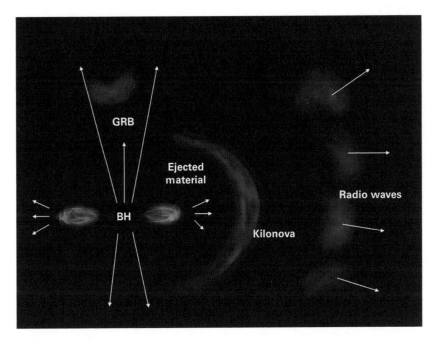

Plate 7

A sketch representing the salient characteristics of a collision of neutron stars, in which the material ejected by the rupture of one of them rotates in a disk at speeds that are considerable fractions of the speed of light. The production of gamma rays and an envelope of light (shown in red) can be seen and is much more isotropic than the emission of gamma rays. An observer looking from the top right would see the gamma ray burst emission (GRB) and the burst of light as well. (Credit: Adapted from figure 3 of B. Metzger, "Kilonovae," *Living Reviews in Relativity* 20:3, 2017.)

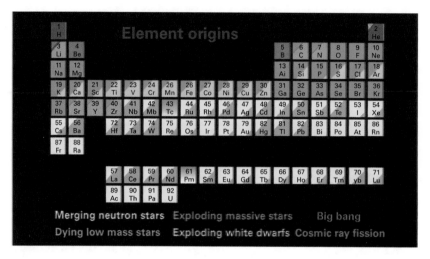

Plate 8

Periodic table of the elements indicating their astrophysical origin. (Credit: Jennifer Johnson/SDSS/CC BY 2.0.)

8
A Brief History of Terrestrial Gravitational Wave Detectors

As Einstein predicted in 1916, the effect of gravitational waves is very small. Nevertheless, modern technology has allowed the LIGO observatories in the US and the Virgo interferometer instrument in Europe to detect gravitational waves from violent astrophysical collisions. This chapter is dedicated to discussing the development of this technology.

8.1 The Small Amplitude of Gravitational Waves

Gravitational waves manifest themselves through changes in distances between any pair of objects or a change in the size of an object itself. For instance, if a wave were to go through a circular ring of particles that are in free fall (floating in space, or falling toward the Earth, for example),[1] the distances would change, transforming the circle into an ellipse, then into a circle again, then into an ellipse in the orthogonal direction, as shown in figure 8.1, and repeating the changes in a periodic fashion. The orientation of the ellipses can be in any direction in the plane of the ellipse, so gravitational waves have two possible polarizations: they are usually

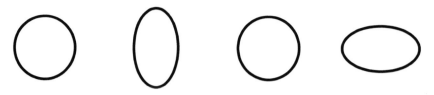

Figure 8.1

Effect of a gravitational wave on a ring of floating particles. The wave travels perpendicularly to the plane of the ring.

referred to as the plus (+) polarization (where the ellipses axes align vertically or horizontally as in a plus sign; see figure 8.1) and the cross (×) polarization (where the ellipses axes seem to form a multiplication sign).

The effect is proportional to the distance that one measures: the amplitude of the wave (or strain) indicates the percentage change in the distances. This has an advantage: the greater the distance between objects, the larger the effect, and the easier it is to measure. The amplitude of gravitational waves is measured as "strain": the change in distance divided by the distance, or the fractional change in distance (it doesn't have units, because we are dividing a distance by another distance).

A gravitational wave going through the Earth not only changes distances, it does so in a very particular way. As shown in figure 8.1, if the distance in a given direction increases between particles on a ring, then in the perpendicular direction, the distance between particles will shorten. These waves interact with everything, in principle making objects (and even the Earth itself) vibrate, but the interaction is so weak (remember the very small amplitudes) that the universe is, for all practical purposes, transparent to gravitational waves! The wave could reach us from below, having traveled through the Earth.

Could a wave be produced and measured inside a laboratory? This is how Heinrich Hertz proved the existence of electromagnetic waves, demonstrating Maxwell's theory of electromagnetism.

A simple source of gravitational waves is a system of two objects rotating around each other. That is, two masses connected by a bar rotating in the plane containing the two masses around an axis perpendicular to that plane, like the propeller on a plane. To produce large amplitudes, we need large masses and high speeds. And as observers, we also need to be not too far from the system that emits the waves (the farther we are, the smaller the amplitude of the waves will be). We could imagine a gravitational wave generator with masses of 2,000 lbs, separated a yard from each other, rotating very fast, let us say at 500 revolutions per second, or 500 Hertz (this would be very difficult in practice, but we are building a thought experiment). The frequency of the gravitational waves produced would be 1,000 Hertz (twice the rotational frequency) and would have a wavelength of 300 km. To measure a true gravitational wave we would have to be a few wavelengths away, let us say just one wavelength away. The amplitude (strain) of the produced wave for this setup at 300 km would be less than one part in billions, or even quadrillions: it would be 10^{-38}, or one part in 10^{38} (a one followed by 38 zeros). Even for large, kilometer-sized instruments, this measurement would be impossible with current technology.

For a stronger source, we need much larger masses, astrophysical ones, but those are farther away. We have already talked about one such example in section 4.7: the two neutron stars rotating around each other in the Hulse–Taylor binary pulsar discovered in our galaxy. But the effect of such waves is still very small, a strain of about 10^{-23}: the distance between the Sun and the Earth would change by about 10^{-12} m, a trillionth of a meter. This system would also produce waves with very low frequency, with a period of 4 hours (half the period of 7.75 hours measured for the binary) which are more difficult to detect (there are many other variables that would change in that time, and therefore contribute to the noise).

But gravitational waves produced by a binary stellar system take energy away from it, so the frequency would increase with time. This energy loss makes the stars become closer, orbiting faster until they eventually fuse into a single object. The wave is therefore produced with an amplitude and frequency that grow with time until the moment of the fusion, after which the amplitude falls off quickly. This waveform (shown in figure 6.1) is known as a "chirp," like the tweet of a bird.

For the final embrace of the neutron stars, in the case of the Hulse–Taylor pulsar, the waves will be 100,000 times bigger than what they are now during their slow waltzing; a massive burst of gravitational waves like this could be measured with terrestrial detectors. However, it will take a long time before convergence occurs: at least about 300 million years, not fast enough even for patient physicists. Although there are other binary systems in our galaxy, it is estimated that about only once every 30,000 years will one of them merge and produce measurable gravitational waves.

Thus we need to consider binary systems close to merger, and those are more likely far away from our galaxy. The first gravitational wave, detected in 2015 and still the largest strain observed at the time of the writing of this book, had an amplitude of 10^{-21}. This stretched and squeezed the 4 km LIGO arms a distance of less than 10^{-17} m. It was produced by two black holes of approximately 30 solar masses each, in a distant point in our universe about 1,300 million light-years from us, orbiting around each other 100 times per second and traveling at half the speed of light when they merged. And in spite of this small amplitude, the gravitational waves emitted were detected! How could LIGO be capable of measuring such minute changes in distance that are much smaller than the particles inside an atom? We are dedicating this chapter to telling the story of this techno-logical feat. It is also a wonderful tale about the work and dedication of the physicists, technicians, and engineers who created LIGO and Virgo and continue working to improve their sensitivities.

8.2   Bar Detectors

The first attempts to detect gravitational waves in the 1960s were not based on a clear understanding of the strength of the sources involved or the expected amplitude of the waves to be detected, but instead focused on building a detector and taking data. Sometimes even if nothing is detected (an experiment with a null result), the knowledge can be used to test the maximum possible strength of astrophysical sources. If there is a positive result, the discovery often requires new explanations and theories. The pioneer in the experimental search for gravitational waves was Joseph Weber, a professor at the University of Maryland.

Weber started his studies at Cooper Union in New York City, but to save money on room and board, he transferred to the US Naval Academy in Annapolis, MD. During the Pearl Harbor attack, he was aboard the aircraft carrier *Lexington*, which was at sea. He survived the sinking of "Lady Lex" in the Coral Sea Battle. At the end of the World War II, he was hired as an electrical engineering professor at the University of Maryland. He became one of the innovators behind the development of the laser and the maser (a precursor of the laser based on microwaves). In fact, he was nominated for the Nobel Prize for the invention of the laser. But in the end, the prize went to the American physicist Charles Townes and the Soviet physicists Nikolay Basov and Alexander Prokhorov in 1964. In 1955, he received a Guggenheim fellowship that allowed him to take a sabbatical at Princeton University with John Archibald Wheeler as his advisor. Wheeler suggested that Weber design an instrument capable of detecting gravitational waves. Weber would dedicate the rest of his life to this project.

The instruments Weber created to detect gravitational waves consisted of aluminum cylinders between 0.5 and 1 m in diameter and 1.5 m in length suspended by metal fibers (see figure 8.2). The idea was that if a gravitational wave passed through them, the

Figure 8.2
Joe Weber with one of his bar detectors. (Credit: AIP Emilio Segré Visual Archives.)

cylinders would vibrate like a a glass resonating when the soprano singer hits the right tone (frequency). For this reason, these detectors were called "resonant bars." Weber even attempted to extend this technology to the Moon. With the Apollo 17 mission, he sent a sort of seismograph, with the idea that when a gravitational wave went through the Moon, it would make it vibrate and that vibration could be detected. Unfortunately, the device was not deployed properly and it did not work.

Weber considered that the effect of the wave might be so small that it could only be confirmed if it was observed in two detectors. This principle of coincidence is fundamental in the history of the field. He also understood that even if there were coincidences, these could be just due to noise in the detectors. He therefore conducted statistical studies that computed the probability that such coincidences were not due to the noise. Weber was the pioneer of this technique, which is used today in all detectors in operation. Weber determined that in his measurements, the noise produced by the motion of the atoms in the bars (what is called "Brownian motion") was one tenth the size of a proton! His bars had a sensitivity of about 10^{-16}, measuring coincident variations in his two instruments (one at the Argonne National Lab in Chicago and the other in Maryland).

In 1969, Weber announced that he had discovered gravitational waves. For comparison, the waves detected by LIGO were 100,000 times smaller in amplitude than the sensitivity achieved by his resonant bars.

The discovery was huge, so many scientists were skeptical. A lot of enthusiasm went into trying to reproduce the results, since the detectors were not prohibitively expensive. Even Stephen Hawking published an article in 1971 with his student Gary Gibbons, proposing better detectors and better methods to analyze the observations and speculating on possible astrophysical sources.

Richard Garwin (an American physicist about whom it is claimed that his supervisor Enrico Fermi said that he was "the only genius he ever met") built a detector similar to Weber's at the IBM laboratories. He failed to make any detection. Similar detectors were built in many countries, but none of them could find gravitational waves. A great controversy ensued, leading to articles pointing out errors and letters to the editor in the magazine *Physics Today* published by the American Institute of Physics (AIP). The general consensus was that Weber had not detected gravitational waves.[2] In spite of all this, the field of gravitational wave detection was born; the challenge was to build more sensitive instruments.

Throughout the years, bars were built in laboratories around the world and grew in complexity and sensitivity. Their thermal vibrations were minimized by cooling them down to very low temperatures using sophisticated cryogenic technologies. Although there were many technological advances over several decades, the minimum noise (maximum sensitivity) that was achieved in the end was 10^{-21} at frequencies around 1,000 Hertz. More sensitive bars designed to operate at lower frequencies were planned, but they would have to be bigger and more expensive. By the time these proposals for new bars were considered, the technology for interferometers had already achieved similar sensitivity with much a bigger potential for improvement. The last bar operating in the US, at Louisiana State University, ceased operations in 2005. Weber passed away in the year 2000, still claiming he had detected gravitational waves. Many of the scientists that built bars at this time were later involved in building interferometric detectors, so the field of experimental gravitational wave astronomy had been born with Weber.

8.3   Interferometric Detectors

Interference is a phenomenon in which two waves add up to form another wave that could have a larger or smaller amplitude. It could be constructive or destructive or anything in between. For instance, if

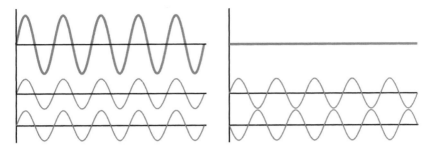

Figure 8.3
Interference of two waves. Left: The two lower waves are in phase and create constructive interference; the resulting wave has twice the amplitude. Right: The two lower waves are 180° out of phase, resulting in destructive interference. (Credit: Haade; Wjh31, Quibik, Wikipedia.)

the crests and valleys of the waves coincide in position, they add up, making a larger wave. The type of resulting interference depends on the relative phase of the superimposed waves, as shown in figure 8.3.

An example of an interferometer is the one used by Albert Michelson and Edward Morley in their experiments to measure the speed of light in the aether in the 1880s. Modern interferometers are optical instruments that typically use the interference of a laser source to measure distances.

In 1962, Mikhail Gertsenshtein and Vladislav Pustovoit in Moscow published a brief paper suggesting the use of a Michelson type interferometer to detect gravitational waves, but the paper did not attract a lot of attention and was largely forgotten. The Soviet scientists were, however, able to present their work at a conference in Copenhagen, and Weber, who also attended, started to consider their suggestion as a possibility.

In the mid-1960s, students at MIT in a class on general relativity that was being taught by Rainer ("Rai") Weiss (at the time a young assistant professor) asked him about Weber's experiment and gravitational waves. Weiss came up with the idea of illustrating the concept by discussing the effects the waves would have on a Michelson interferometer in which the mirrors were suspended by pendulums ("free" masses in the horizontal direction), allowing them to move when the gravitational waves pass through them. He kept on thinking about the idea and built a prototype. In 1972, he wrote an MIT internal report for the Research Laboratory of Electronics, which was funding Weiss's research on cosmology and gravitation. The report was coauthored with one of Weiss's students, among others. In its 23 pages, their report discussed for the first time all the noise sources present in an interferometric detector and argued that it was possible, technologically, to get the noise levels low enough to attempt the detection of the sources that were considered by Weber or other sources. Weiss had invented LIGO. He was skeptical, as many others had been, about Weber's detections, but he described pulsars as a plausible source of gravitational waves, although the sensitivity would require kilometer-sized detectors. For a while, Weiss moved on to other topics. In particular, he took part in the measurement of the cosmic microwave background radiation by the COBE satellite.

Weiss started testing the concept of using an interferometer to measure strain with a prototype with arms 1.5 m long, recognizing this as a prototype test for detectors that needed to be much larger. Progress on this was slowed down due to the Mansfield amendment mentioned in section 4.6, since the Research Laboratory of Electronics received financial support from the military and so adopted the same policies. The experiment was instead funded by NSF beginning in 1975. The first interferometer intended to detect gravitational waves was built in the late 1970s by a former student of Weber, Robert Forward, who published a paper on a search for gravitational waves using the device in *Physical Review*, where he acknowledged advice from Rai Weiss.

Based on Weiss's ideas, research on interferometric gravitational wave detectors was started in European labs, too. In Europe in the 1970s, two groups started work on the topic. One of them was led by Heinz Billing at the Max Planck Institute in Garching, near Munich, Germany and the other by Ron Drever in Glasgow, Scotland. In 1975, the Max Planck group started research on a 3 m long interferometer. This prototype paved the way for the construction of a larger one (30 m) in 1983. In 1977, a 10 m prototype was built in Glasgow. In 1985, the Garching group proposed the construction of a large detector with 3 km arm lengths. The Glasgow group proposed a similar instrument in 1986, and these groups combined their efforts in 1989. Thus the GEO project was born, which in the end, due to budget restrictions, became an interferometer 600 m in length (and then called "GEO600"). It was built in Sarstedt, south of Hanover in northern Germany. The construction of GEO600 started in 1995. This interferometer played and still plays an important role as a prototype for the development of technologies for present-day interferometers.

While these initiatives were under development in Europe, during a chance meeting at a conference in the US in 1975, Weiss and Thorne exchanged ideas about the future of the detection of gravitational waves. Thorne became convinced that a large-scale interferometer had to be built. He was already quite an influential scientist—he was elected to the US National Academy of Sciences when he was 33 years of age. He convinced Caltech to create an experimental group, hiring Ron Drever from Glasgow and Stan Whitcomb from Chicago (an American scientist who had just finished his PhD in astronomy).

In 1979, NSF approved funding for the construction of a 40 m long prototype interferometer at Caltech, led by Drever and Whitcomb. This and the prototype at MIT demonstrated that the technology required to control instrumental noise at levels sensitive enough to detect gravitational waves was within reach. The effect of the waves is larger the longer the interferometer. So the next step was the construction of a larger interferometer at a scale big enough to secure success. A key 1983 report at MIT, with the collaboration of the Caltech group (commonly known as the "blue book") made a realistic study of a long baseline gravitational wave "antenna system" including a vacuum system. The report was coauthored by Paul Linsay, Peter Saulson, Rainer Weiss, and Stan Whitcomb together with industrial consultants.

In 1986, NSF created an evaluation committee of experts, chaired by Andrew Sessler, that studied the situation in detail. This commission recommended enthusiastically the construction of a large-scale interferometer, which would later be called the LIGO project (Laser Interferometer Gravitational-Wave Observatory), but with a more formal administrative structure. In 1987, Rochus ("Robbie") Vogt, a Caltech professor, was named LIGO director. Together with Drever, Thorne, Weiss, and Fred Raab—who had arrived from the University of Washington—they sent a proposal to NSF to fund the construction of two interferometers of 4 km in length. The proposal was to build two initial detectors that might not have much chance of detecting gravitational waves but could later be upgraded to an advanced configuration that would. The reason for the two-step approach is that technologies that are tested in the lab do not necessarily scale linearly with increased size. The large size was thought to allow the development of instruments of increasing complexity.

In 1991, the US Congress included LIGO in the national budget, but the definitive funding was still pending due to requirements from the National Science Board—the board that oversees NSF—and some additional problems with the project. In 1992, two sites were chosen, one in Hanford, Washington, and the other one in Livingston, Louisiana (figure 8.4). The project followed the principle established by Weber of confirming a detection by observing it separately in two detectors far away from each other.

Approval by the US Congress was by no means guaranteed. Congress seeks input from the community about what it approves. In this case, there was great opposition by US astronomers. Astronomers tend to have a very conservative attitude toward new projects. In particular, when a new instrument is proposed, the proposal has to specify what is going to be observed, and in the case of LIGO, there were large uncertainties about that. The instruments were

Figure 8.4
Aerial views of the LIGO observatories in Livingston, Louisiana (left panel), and Hanford, Washington (right panel). (Credit: Caltech/MIT/LIGO Lab.)

expected to observe collisions of binary systems of compact objects, but it was not known how frequently such collisions would take place. The only known systems at that time were three binary pulsars that were not expected to merge for several hundred million years. Trying to extrapolate an estimation of the number of binary neutron star collisions per year that one may expect in a given galaxy from this extremely limited known population was bound to be a bold conjecture. And at that time, not even one binary black hole system had been identified. The abovementioned conservative practice of the astronomers has, paradoxically, led to many surprises. For instance, when the Mt. Wilson telescope in Los Angeles was proposed—at the time the largest telescope in the world—it was expected that it would help study properties of stars whose brightness varies with time known as Cepheids. But eventually, Edwin Hubble ended up using these Cepheids to estimate the distances to galaxies and discovered the expansion rate of the universe.

In an April 30, 1991, article in the *New York Times*, the doubts of the astronomers were summarized. A famous astronomer—at the time at AT&T Bell Laboratories—Anthony ("Tony") Tyson is quoted mentioning that he had conducted a poll among astronomers with a result of 4 to 1 against the construction of LIGO. The poll was prepared when Tyson—who had devoted some effort to detecting gravitational waves with bars in previous years—testified in front of the Subcommitteee of Science of the Commision of Science, Space and Technology of the US House of Representatives, when LIGO's funding was discussed.[3] Although the objective of that testimony was to oppose LIGO financing, its description of the smallness of the amplitude of gravitational waves is correct. Tyson said:

> Imagine this distance: travel around the world 100 billion times (a total of 2400 trillion miles, or one million times the distance to Neptune). Take two points separated by this total distance. Then a *strong* gravitational wave will briefly change that distance by less than the thickness of a human hair. We have perhaps less than a few tenths of a second to perform this measurement. And we don't know if this infinitesimal event will come next month, next year, or perhaps in thirty years.

But the presence in NSF of Richard Isaacson (the expert on gravitational waves discussed in chapter 4) was crucial. Isaacson and Walter Massey (the director of NSF at the time) made several clever moves that ensured that funding was approved. In particular, it is rumored (these things are not public) that among the many proposed sites, the election of the Louisiana site was to take advantage of its legendary politicians, historically disproportionately influential for a small state. One of the strategies was not to include the cost of LIGO in the NSF budget but assign it a separate line item in the US budget. That made it clear that it was a one-time payment and not an expansion of the agency, which would have caused the opposition of conservative politicians.

The project had some initial problems. It is a very different experience to lead a scientific experiment in a small academic laboratory than to manage a megaproject like LIGO. Its management requires different abilities, more akin to those of a corporation CEO than those of a scientist. One of the indispensable skills is to manage personnel—in particular, scientists—who are in general very independent spirits. This had already been observed during the Manhattan

Project, when the US Army tried to manage scientists like Richard Feynman, who continually played practical jokes on the military rules, which was not appreciated by the brass.

Vogt knew how to run large projects. He had worked on the NASA Voyager mission, had been director of the Division of Physics, Mathematics and Astronomy at Caltech and director of the radio telescope that Caltech has at Owens Valley. Despite his experience, he clashed with the creative and chaotic spirit of Ron Drever. When the latter traveled in 1992 to the thirteenth International Conference on General Relativity and Gravitation in Huerta Grande, near Córdoba, Argentina, he found upon his return to California that the locks in his lab had been changed. Apparently it was all a misunderstanding, but the scandal reached the pages of the *Los Angeles Times*, which titled the piece "Leggo my LIGO."

In the end, Vogt was replaced as LIGO Executive Director with another Caltech professor, Barry Barish, a particle experimentalist with experience managing large projects. Barish had worked on the ill-fated Superconducting Super Collider (SSC) in Texas, which was going to be the largest particle accelerator in the world, but the US Congress had canceled it due to cost overruns. Initially budgeted at $4 billion, the costs had soared to more than $12 billion. This type of accelerator was finally built in Geneva, Switzerland, at the European Center for Nuclear Research (known as CERN, for its initials in French). It ended up discovering the "god particle," the Higgs boson, in 2012. Together with Barish came other former SSC scientists and engineers who played crucial roles in the initial development of LIGO.

Another problem of a political nature arose. In 1994, during Bill Clinton's presidency, the Republicans made major gains in the midterm elections, obtaining control of both houses of Congress for the first time in more than 50 years. The Republican platform, known as "Contract with America," included cutting government spending. The Speaker of the House and architect of the "Republican Revolution of 1994" was Georgia congressman Newt Gingrich. At NSF, Isaacson and Neal Lane—the NSF director at the time and future science advisor of Bill Clinton—feared the worst: LIGO was a low hanging fruit, as it is easy to cancel a project that has not yet started. Isaacson and Lane asked for a meeting with Gingrich to discuss the subject, and although they said that they were shaking as they entered the room, they were very surprised by what happened. It turns out Gingrich is a "techie," and he loved the project and assured them that LIGO would not be cut. And it was not cut.

The budget was updated with $270 million for construction, constituting the biggest project funded by NSF up to that time from a new account (Major Research Equipment and Facilities Construction). The account has since been used for other major projects, including neutrino searches in the Antarctica (IceCube) and a large survey of space and time with the Vera C. Rubin Observatory being built in Chile. The cooperative agreement for LIGO in the period 1992–2003 amounted to $360 million, including money for operating the observatories.

The construction of the buildings housing the interferometers started in 1994 and was finished in 1997. The installation of the instrumental configuration of what was called "Initial LIGO" was finally finished in 2001. It started gathering scientific data through its first scientific run, denominated S1, in August 2002. The detectors' sensitivity was improved progressively, obtaining scientific results that we will detail in chapter 10. In 2005, scientific run S5 started and continued through 2007 achieving 1 year of data gathering at the original design sensitivity planned for Initial LIGO. These projects take time: 13 years passed from ground breaking through the completion of the original scientific goals of the project.

By 1997, the LIGO project was rearranged around two main interconnected structures: (a) the LIGO Laboratory ("LIGO Lab"), consisting of the personnel at Caltech, MIT, Livingston, and Hanford, responsible for the operation of the detectors; and (b) the LIGO

Scientific Collaboration (LSC), responsible for the organization of the research and development of new technologies, coordination of the data analyses necessary to search for gravitational waves, writing up results for publication, and eventually validating detections. The LIGO Laboratory personnel were also members of the LSC. The group of researchers working on GEO600 integrated themselves into the LSC.

The LIGO Lab is managed by the executive director of the project. The LSC is led by elected spokespersons with 2-year terms. Weiss was the first spokesperson starting in 1997, followed in 2003 by Peter Saulson of Syracuse University, the first formally elected spokesperson; in 2007 by David Reitze of the University of Florida; in 2011 by Gabriela González of Louisiana State University, who served as spokesperson for three terms; in 2017 by David Shoemaker of MIT and Laura Cadonati from Georgia Tech; and in 2019 by Patrick Brady of the University of Wisconsin-Milwaukee, the current spokesperson (2022). The LSC currently has over 1,400 scientific members from more than 100 scientific and academic institutions in 20 countries.

In parallel with the LIGO project in the US, a French–Italian collaboration established the Virgo project, a 3 km interferometer in Cascina, near Pisa, Italy. The LIGO and Virgo Scientific Collaborations signed a cooperation agreement in 2007 and analyze their data jointly. The Virgo project has many similarities with LIGO, but it introduced new technologies, particularly in the suspension of its mirrors. It had multiple suspensions from the beginning, something that LIGO only acquired in its Advanced version. The sensitivity goal of Virgo is fairly similar to that of LIGO, but an important difference is the length of the interferometer arms, which at 3 km are 1 km shorter than LIGO's arms. Virgo's funding was delayed for a few years with respect to that of LIGO, which is the reason its progress in achieving sensitivity was lagging LIGO's. Over time, participants from a dozen other European countries, as well as China and Japan, joined the Virgo Collaboration.

In 2005, after several years of hard work, the Initial LIGO detectors achieved the sensitivity that had been designed for at frequencies above 80 Hz (sensitivity at low frequencies is always more challenging, as we will discuss in chapter 9). Although no gravitational waves were detected in 2 years of observations in 2005–2007, the quality of the data obtained resulted in the publication of significant astrophysical upper-limit estimations on gravitational wave emission and delivered clear proof of the technological concept. In 2008, the National Science Board approved the funding of the second phase (Advanced LIGO), with the goal of achieving a 10 times better sensitivity, which began to be installed in 2010.

By 2015, the instrument had already a sensitivity 3–4 times better than the initial interferometers, and a campaign to obtain good quality data for a few months was undertaken, before continuing more work on improving the sensitivity. There were only low expectations of detecting gravitational waves, at least initially. However, on September 14, 2015, shortly after starting to take data, the first detection took place. Curiously, on that day and approximately at the same time, Albert Einstein's apartment in Bern, where he lived while working at the patent office, was being designated a historic site jointly by the European and the American Physical Societies.

In 2017, Barry Barish, Kip Thorne, and Rainer Weiss were awarded the Nobel Prize in Physics, for "decisive contributions to the LIGO detector and the observation of gravitational waves." Fittingly, the affiliations noted for them in the press release of the Royal Swedish Academy of Sciences were not Caltech/MIT, but "LIGO/Virgo Collaboration," a recognition of the hard work of hundreds of scientists and engineers who made the discovery possible.

By the time of the discovery in 2015, the total US Federal Government investment in the LIGO project had exceeded $1.1 billion. This was a high-risk, potentially high-yield bet that NSF took, and it paid off handsomely. NSF's authorities have emphasized that perhaps that is exactly the type of basic science a government agency like it should undertake, because no one else would.

9
The Technology of LIGO

Using an interferometer to measure changes in distances is not a new technology: Michelson and Morley had already done it in 1887 in their famous experiment to measure variations of the speed of light with respect to the aether. What is new in gravitational wave detectors is their size and the resulting sensitivity. Initial LIGO was a trillion times more sensitive than Michelson's interferometer, and Advanced LIGO is already several times better than that: it can resolve changes in length as small as 1/10,000 of the radius of a proton, over a length of 4 km. Amazing! How is this done?

9.1 Interferometry: Differences of Length

The key idea is to use a laser light beam and split it in two using a semi-transparent mirror ("beam splitter") at an angle of 45 degrees with respect to the incident beam. This allows half of the light beam to be transmitted and reflects the other half, resulting in two beams traveling in perpendicular directions: the arms of the detector, which are 4 km long. The two beams travel this distance to bounce off mirrors placed at the end of each arm. And the bouncing beams interfere when they meet back again at the original point.

If the two distances traveled by the beams are identical, the waves emerging from the detector interfere negatively or "destructively" (the crests of one wave coincide with the valleys of the other, and so they cancel out each other, as shown in figure 8.3 in chapter 8). No light comes out of the interferometer output port! Obviously, no magic is involved: the waves that go back to the laser interfere constructively, and the interferometer behaves like a mirror in that sense.

If the lengths of the arms are different, the destructive interference is not complete, and some light comes out, depending on the difference between the lengths of the two paths traveled by both beams. The intensity and the variation with time of the light coming out is measured with a photocell at the interferometer's output. This change in intensity measures the change in arm lengths. Figure 9.1 depicts two possible situations. The top panel depicts when both arms are of equal length. In that case, the waves are out of phase, and the light cancels at the photocell positioned at the lower right in the figure (destructive interference). The bottom panel shows different lengths for each arm. In that case, some light reaches the photocell. If the waves are in phase (constructive interference), then all the light exits the interferometer, and no light comes back toward the laser. The relation between the valleys and crests of the wave and the shortening and lengthening of the distance between the mirrors in the interferometer can be seen in figure 9.2.

Why is a laser used? An ordinary light source, like an electric bulb, emits a superposition of electromagnetic waves with very different wavelengths. A typical laser source emits *coherent* light—light that has the same wavelength and is emitted in phase. This makes interferometry easier, although interferometers can also be built with ordinary light, as was the case for Michelson and Morley. LIGO uses an infrared laser, invisible to the human eye, with a wavelength of 1 micron (a millionth of a meter). The detector can measure changes in the interference pattern due to differences in phase smaller than 1 part in 10 billion.

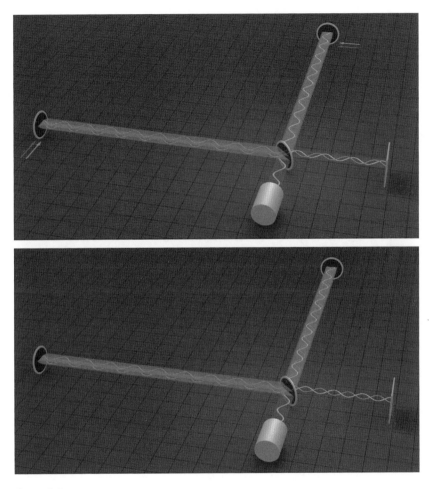

Figure 9.1
How a Michelson interferometer works. See explanation of the images in the text.
Note that the motion of the mirrors is tiny and is indicated by the separation of the
arrows in the bottom panel. (Credit: LIGO/T. Pyle.)

The measurement of the amplitude of a gravitational wave (the
strain) is the distance the arms will be stretched or squeezed by the
wave, divided by the 4 km length of the arm. However, not only
gravitational waves can produce changes in the arms' lengths. There
will also be a signal at the output when the mirrors move, when the
atoms of the mirror vibrate (which they always do!), when the laser
beam does not travel in a straight line, or for many other reasons.

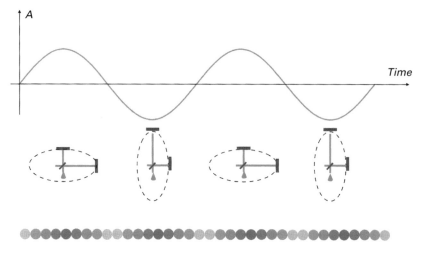

Figure 9.2
Effect of a very simple gravitational wave (with its amplitude A plotted as a function of time) on the distances between the mirrors at the end of the interferometer. It can be seen how the crests and valleys of the wave coincide with the squeezing and stretching of the distance between the mirrors. The shades of the circles indicate the intensity of the light at the photocell.

The effect of all possible disturbances, other than a gravitational wave, that will cause the interferometer arms length to change is called the detector "noise." LIGO's amazing design and its scientists' and engineers' unrelenting efforts have made this noise small enough so that the instrument is able to detect tiny changes in length produced by astrophysical sources. The smaller the noise is, the smaller will be the gravitational waves that LIGO can detect.

The goal of Initial LIGO was to prove that sensitivity to distance changes already achieved in laboratory prototypes and in Weber bars was also achievable with kilometer-long arms. With that sensitivity, the capability to detect the minuscule amplitudes of gravitational wave amplitudes would be greatly increased, since these changes would be divided by kilometer lengths instead of meters. This would be a factor of 100 improvement with respect to the longest prototype at Caltech, which was only 40 m long. Also, prototypes could achieve low levels of noise at frequencies of thousands of Hertz (kiloHertz, or kHz), but astrophysical signals exist at lower frequencies of around 100 Hz. The design for Initial LIGO was to measure noise smaller than 10^{-21} in strain between 50 Hz and 2 kHz, suitable for detecting collisions of stellar-mass black holes, of neutron stars and supernovae explosions. The fraction 10^{-21} is like comparing a hair width with the distance to the nearest star, Proxima Centauri.

This sensitivity level would be enough to detect mergers of binary neutron stars at 45 million light-years: the distance to the Virgo cluster, the group of galaxies closest to Earth. Although when construction started, not much was known about how many neutron star binaries existed in each galaxy, it was estimated that at these distances, these events would occur once every 50 years. This is a very depressing number for an experiment, but many times astronomical estimations can be wrong by orders of magnitude. Maybe it happens every 5 years, and the experiment could get lucky! In any event, what drove the researchers who developed and implemented the technology involved was the excitement of achieving these incredible levels of precision. There was also better technology being developed for a second phase of the project, Advanced LIGO, which would improve the sensitivity by a factor of 10.

9.2 No Air Allowed: Vacuum System

The laser beam in the detectors needs to travel unimpeded to the mirrors and back. Even minute particles or loose floating atoms would deflect the straight path of the light, create noise, and diminish the instrument's sensitivity. For this reason, the light travels through tubes. Ultrahigh vacuum pumps are used to keep the tubes free of air or gases of any kind (only a trillionth of atmospheric pressure is left inside). Being 1.2 m in diameter and 4 km long, LIGO tubes are among the largest ultrahigh vacuum installations in the world, comparable to the Large Hadron Collider (LHC, the most poweful particle accelerator in the world, managed by the European laboratory CERN in Switzerland), which is longer than LIGO, but narrower. This vacuum installation is one of the biggest costs in the construction of the detectors and it has to be done at the outset; it cannot be added later as an improvement. To create the ultrahigh vacuum in the tubes, 20 m long pieces of the tubes were manufactured in a dedicated facility near each observatory and were welded together on site. Afterward, the insides of the tubes were carefully cleaned, and then the stainless steel was heated so that molecules of air and other substances from the surface of the tubes were expelled from them and then trapped and extracted by the vacuum pumps. The 8 km of tubes at each observatory were covered with thermal blankets powered by electrical generating trucks to essentially bake them and eliminate any residual gases in them. The extra portable vacuum pumps operated for 40 days to achieve the necessary vacuum in the system. (The high temperature was applied for 30 of those days—a hefty electrical bill!) The vacuum in the system is maintained by six permanent ion pumps at the corner station and the end stations (see figure 9.3).

To avoid the need to repeat the pumping process every time an upgrade or any work is needed on the subsystems, special valves were installed along the vacuum tubes and at the chambers enclosing the mirrors and suspension systems. Thanks to these valves, the vacuum inside the tubes has been maintained since finishing the installation in 1995 throughout the different upgrades and improvements during LIGO's history. This vacuum is always carefully monitored. In 2008, at the Louisiana site, some small leaks in one of the valves were detected, and it had to undergo a high-risk but ultimately successful repair. Other minor leaks at the same site have been associated with corrosion due to microbes that seem to be prevalent in

Figure 9.3
The vacuum chambers at the end of the tubes. (Credit: LIGO.)

the southeastern US. The microbes are carried by animals (rats!) that crawled inside the thermal blankets that were left around the tubes. The blankets have been removed, and dehumidification efforts help the situation, but animals keep bringing the microbes.

The vacuum tubes are located inside a protective tubular concrete structure about 3 m in diameter with access doors for service. This structure has protected the vacuum tubes from alligators, hurricanes, hunters' bullets, crashing cars, and brush fires—all of which have actually happened! Once a group of hunters in the woods around the Livingston Observatory shot at one of the buildings at the end of the arms. The concrete stopped the bullets, and eventually, a friendly barbecue with the local hunting club put an end to the shots. On the other coast, at the Hanford site, a brush fire ran over the detector but without causing damages. At the same site, the local police used the land around the interferometer to practice car chases. Figure 9.4 shows how one of their vehicles crashed into the external concrete enclosure, fortunately with no damage to the facility and only minor injuries to the driver. The Livingston Observatory has closed several times during hurricane season, and several

Figure 9.4
Police car crashes against the Hanford vacuum enclosure. (Credit: Caltech/MIT/LIGO Lab.)

hurricanes have impacted the site, but the storm damage has been minimal.

The length of the interferometers requires taking into account the curvature of the Earth to keep the beam tubes straight. This was another of the engineering challenges: a straight concrete slab had to be created to mount the tubes, which required correcting elevations of up to 3 feet. Corner stations are mounted on concrete slabs separated from the walls by rubber dividers, which minimizes motion of the walls due to wind that would otherwise influence the detector.

9.3   Suspended Mirrors

The changes in distances produced by gravitational waves must be measured between masses that are "free" of any other forces. In the

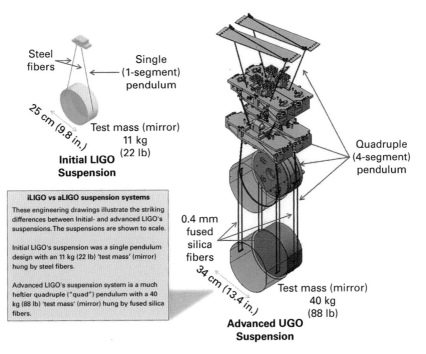

Steel fibers

Single (1-segment) pendulum

25 cm (9.8 in.)

Test mass (mirror) 11 kg (22 lb)

Initial LIGO Suspension

Quadruple (4-segment) pendulum

iLIGO vs aLIGO suspension systems

These engineering drawings illustrate the striking differences between Initial- and advanced LIGO's suspensions. The suspensions are shown to scale.

Initial LIGO's suspension was a single pendulum design with an 11 kg (22 lb) 'test mass' (mirror) hung by steel fibers.

Advanced LIGO's suspension system is a much heftier quadruple ("quad") pendulum with a 40 kg (88 lb) 'test mass' (mirror) hung by fused silica fibers.

0.4 mm fused silica fibers

34 cm (13.4 in.)

Test mass (mirror) 40 kg (88 lb)

Advanced UGO Suspension

Figure 9.5

The Initial LIGO suspension system (a simple pendulum) and the Advanced LIGO quadruple suspension system for test masses. (Credit: LIGO.)

interferometric detector, the mirrors are the "free masses" (as shown in figure 9.5). For laboratory interferometers, the mirrors were bolted to optical tables. But the mirrors cannot be completely free, because they would then fall to Earth, and the experiment would be very short lived. The LIGO mirrors at the end of the 4 km arms are suspended as pendulums: this lets them move (almost) freely in the horizontal direction of the arms' length, but it does not let them fall. Moreover, this suspension has an additional benefit: although the frame from which the mirror is suspended can move, at the frequencies of interest, this effect is minimal.

When holding a pendulum from its suspension point, if the point is moved horizontally, the pendulum will also move. But the amplitude of the motion of the pendulum bob will depend on the frequency at which the support point moves. If it is moved slowly (at low frequency), the pendulum bob moves just like the suspension point. However, if the suspension point oscillates rapidly back and forth, the pendulum bob will move much less (try this at home with a yo-yo). There is a special frequency that divides the low and high frequencies, determined solely by the length of the pendulum (not its mass). That frequency is called the "resonant" frequency, because if the suspension point moves at that frequency, the pendulum will move more than the suspension point does.

Therefore, suspending the mirror like a pendulum helps diminish the noise at the desired 100 Hz if the resonant frequency is lower. The pendulums are about a foot long, and the resonant frequency is about 1 Hz, so at 100 Hz, the motion is 10,000 times lower than the suspension point (it goes as the square of the frequency). In Initial LIGO, the mirrors were suspended in simple pendulums like the ones described; Virgo instead used multiple pendulums—mirrors hanging from masses that were hanging from other pendulums—and that was much better (although more complicated). For Advanced LIGO, the mirrors are suspended in quadruple pendulums. The mirrors hang from cylindrical masses of similar size to them, which themselves hang from cantilevers that can vibrate and are held at one end. The complete system is hung from other cantilevers attached to a metal structure (see figure 9.5). Multiple pendulums have a disadvantage: since the resonant frequencies need to be low, the systems become very tall. And all these suspensions need to be inside the vacuum. The vacuum chambers used in Virgo are very tall towers that are more than two stories high!

The sophisticated suspensions used in Advanced LIGO were designed by the LIGO Collaboration group at the University of Glasgow, a group created by Ron Drever. The current leaders are now UK scientists Professor Sir James Hough and Professor Sheila Rowan (who until 2021 was the Scottish Government's Chief Scientific Adviser and is currently the President of the UK Institute of Physics).

These suspensions have to isolate the test mass from ground motion and additionally need to push it so that the detector is kept operational. But they are also required to lose as little energy as possible when vibrating, to reduce thermal noise. Many years of research went into the design and testing of these systems—this is a fruitful example of the value of international collaboration for technological development.

9.4   Seismic Isolation

The surface of the Earth is always moving, not just during earthquakes. Even with the multiple pendulums described, this motion would also shake the mirrors of the interferometer, producing strain noise much larger than 10^{-21}. This seismic noise also needs to be reduced: the technology to do this is called "seismic isolation." An advanced seismic isolation technology had to be developed to eliminate even microscopic quivering of the mirrors.

This technology is different than the one used to construct buildings capable of surviving earthquakes. There is little that can be done about such violent ground shaking, except implementing systems to prevent the mirrors from breaking. Fortunately, strong earthquakes are rare at the LIGO sites, although large earthquakes anywhere around the world shake the LIGO mirrors—the earthquake safety systems have been proven useful more than once! The problem is the regular and imperceptible shaking of the ground. We do not notice it, but the Earth's surface is always vibrating, from some fractions of a micron to a few microns, depending on the location and the frequency considered. The vibration of the ground is maximum with periods of several seconds due to surf waves on the coasts, a frequency much lower than the gravitational wave detector band. Because the surf waves on the coast are weather dependent, so are these effects on the LIGO system. The suspended masses provide much of the needed isolation, but less so at low frequencies. Thus a few more orders of magnitude in reduction of the seismic noise are needed.

The seismic reduction technology in Initial LIGO used one of the simplest physics systems: masses and springs. A mass-spring system is what makes a car suspension work. Automobile factories insert springs in a car to isolate its cabin from the jouncing of the wheels caused by the road's bumpiness. As in the case of the pendulum isolating its bob from the motion of the suspension point, a car suspension's ability to isolate the riders from the road is frequency dependent. Driving over cobblestones, where the wheels go up and down quickly, the car is well isolated from the motion of the wheels. If a car is instead slowly driving on a road that rises and drops, the cabin will move with the wheels, and there will be no isolation from their motion. In this case, the "noise" is below the "resonant frequency" of the suspension.

Initial LIGO used four springs between heavy metal layers, with resonant frequencies smaller than 10 Hz. At 100 Hz, the mirrors moved much less as a consequence. But these values were not good enough for Advanced LIGO, where a sensitivity 10 times higher was needed. An "active" system was required, which monitors electronically the motion and cancels it through suitable devices. The process is similar to the technology in noise-canceling headphones that block environmental noise by monitoring it and using electronics to cancel it out before the sound enters the ear of the user. This active system posed technological challenges for LIGO that were finally overcome. Among these challenges was the use of electronic components inside a vacuum without contaminating the vacuum with the gases that the components emit. An advantage was that the systems could be optimized and improved without making drastic changes in the hardware. In hindsight, the decision to use active systems is very obvious: systems are more compact than equivalent passive ones and are also more flexible. However, at the time the systems were being developed, this was not obvious at all. There were two teams proposing passive systems and active systems for Advanced LIGO seismic isolation, led by Riccardo DiSalvo (an Italian scientist, at the time at Caltech) and Joseph Giaime (Louisiana State University professor and current head of the Livingston Observatory), respectively. There were many drawings, experiments with prototypes, and review committees. Finally, the active system was chosen. The systems have been much improved since then by the LIGO seismic team, including Louisiana State and Stanford scientists. Figure 9.6 shows the active seismic isolation system installed in a vacuum chamber in LIGO, before closing the chamber.

The Livingston site presented additional challenges. The site is closer to the coast and the land in Louisiana is softer. This generated a micro-seismic peak at 0.1 Hz that was so large that it prevented the interferometer from operating, even though it was below the desired frequency band. In addition to this problem, the site is in the middle of a forest that is actively logged. This human activity generated additional noise at around 3 Hz that further complicated operations, especially during the day. To have Initial LIGO work at Livingston, it

Figure 9.6
Left: The active seismic isolation system that the quadruple suspensions hang from.
Right: A test mass installed in LIGO. (Credit: LIGO.)

was necessary to introduce an additional hydraulic noise reduction system. This is an example of what Initial LIGO was meant for: it illustrated the unexpected challenges of going to a kilometer-sized interferometer and the new choice of technologies needed to overcome them. It also illustrated the need to develop techniques for future detectors, as they may become needed in earlier ones.

9.5   A Quantum Enemy: Shot Noise

Another source of noise that limits the astrophysical reach of the detectors is the quantum noise associated with the laser light that is used. According to quantum physics, every wave is also made of particles: this is called the "wave-particle" duality. The particles associated with electromagnetic waves are photons. Quantum mechanics also tells us that the number of photons in a beam cannot be determined exactly, nor can the interval between photons. This is the "uncertainty principle" of quantum mechanics (first formulated by Werner Heisenberg) that also precludes two physical properties (like the speed and position) of a particle from being determined simultaneously. The photocell at the end of the interferometers is precisely counting photons to measure the phase difference of the light returning from the arms. But the number of photons, and the time of their arrival, cannot be exactly measured—they are a "quantum" uncertainty. The resulting uncertainty in the phase measurement is called "shot" noise, because it is like the noise that rain makes on a tin roof. The photons hitting the photocell play the role of rain drops on the roof. The phase noise can be reduced using more photons, increasing the laser power. That is why Advanced LIGO uses a very powerful laser, which was developed by collaborators in Germany, showing again the power of international collaboration.

The choice of laser to be used generated some discussions in Initial LIGO. In principle, since what is measured is the difference between waves, the shorter the wavelength is, the greater the precision will be. A green laser would be better than an infrared one, since it has a shorter wavelength. However, the power and stability depend a lot on the device that produces the laser light. In the 1990s, the most common lasers used noble gases like argon (which produces green light), but the more powerful they were, the more power they needed and the hotter they ran. The resultant heat required liquid refrigeration, which in turn introduces noise through vibrations. However, at about that time, the commercial production of solid-state lasers began. These new lasers had similar power but took less energy to run, were more stable, and offered more room for improvement. But they produced infrared (not green) light. Finally, the decision was made to use the infrared laser, which turned out to be a wise choice, because the solid-state technology proved to be superior in the long run. The solid-state laser used is made of a synthetic crystalline material of the garnet group—a type of silicate—called YAG (Y for ytrium, A for aluminum, and G for garnet).

A 10 watt laser was used at the beginning of Initial LIGO; the final data-taking campaign with Initial LIGO in 2010 used 20 watts. For Advanced LIGO, the original plan was to amplify the laser power to 200 watts. However, in 2013–2014, a more conservative strategy was taken, since the extra power reduces the quantum noise, but it presents many other challenges. As the power is increased almost a hundred times in the beams on the test masses, each mirror gets hotter at its center, and the surface changes its curvature. The deformation can be compensated by heating the mirror rims with an auxiliary laser. But vibrations in the test mass are also amplified by the laser beam, and imperfections in the mirror get worse when additionally heated, scattering some of the light. Although the Advanced LIGO laser can deliver much higher power, 20 watts of power (same as initial LIGO) were used in 2015. The power was increased to 40 watts in 2019.

9.6   A Quantum Friend: Squeezing the Vacuum

We have seen that shot noise can be reduced by using more photons (a more powerful laser), but the laser light amplitude also introduces a quantum uncertainty (noise!), and the noisy light pushes the mirrors, changing the distances to be measured. This noise is called "radiation pressure" and is the evil cousin of "shot noise": if the shot noise is decreased, increasing the number of photons, the radiation pressure noise will increase. The effect of radiation pressure can be reduced using heavier masses: initial LIGO used 10 kg mirrors, Advanced LIGO uses masses 4 times heavier.

It turns out that this description of the quantum noise as the addition of radiation pressure and shot noise is also not quite right. The correct description was only understood in the 1990s by Carlton Caves, then a graduate student of Kip Thorne, who proposed to inject squeezed light in the interferometers to improve their sensitivity. This proposition led to advances in the technology to reduce quantum noise.

If the arm lengths of the interferometer are identical, there is destructive interference at the output, and no light comes out (all the light goes back toward the laser). There are no photons to count, there is only vacuum. But in quantum mechanics, vacuum does not mean empty. And it is the vacuum noise that we measure in the photocell. The vacuum noise has equal magnitude in amplitude and phase. Phase noise is what we called "shot noise," and amplitude noise is what we called "radiation pressure." If more photons are used, both amplitude and phase noise grow, but the signal measured (in phase) grows even more, and the signal-to-noise ratio of the phase difference in the returning beam is increased.

The product of the uncertainties in measuring the amplitude and the phase of the signal is constant and determined by quantum mechanics. If one is decreased, the other will automatically increase. This effect can be used to "squeeze" the noise in phase at the cost of increasing the noise in amplitude. Of course, a price is paid, but this can be done without a more powerful laser. Consequently, no extra heating of the mirrors will occur.

How is the vacuum actually squeezed? A different laser (synchronized with the main laser beam) is used to produce a beam in which the phase noise is squeezed. Its beam is injected into the detector through the output port, and then its amplitude is reduced to zero: the previously existing vacuum is now replaced with a squeezed vacuum.

This technique was developed rapidly in laboratories in the US, Germany, and Australia. It was tested and successfully used in the GEO600 detector in Germany, was briefly successfully tested in Initial LIGO at Hanford in 2010 just before turning it off, and was implemented in Advanced LIGO in its third Observing run in 2019–2020. The technique reduced the quantum noise without increasing the laser power.

Scientists can (and will) do even better: the amplitude noise is more important at low frequencies, because the force due to radiation pressure produces less displacement of the masses at higher frequencies. With the same force, it's easier to move a mass back and forth if it is done slowly rather than quickly. At present, this noise does not yet affect the performance of the LIGO detectors, because there are many other sources of noise at those frequencies, but it will soon be a limiting factor. However, the injected "squeezed" vacuum can be modulated to different frequencies if the beam from the auxiliary laser travels in another long optical cavity before being injected into the output port: the phase noise could be "squeezed" at high frequencies, and the amplitude noise at low frequencies (see figure 9.7). This is called "frequency dependent squeezing," and new cavities are being built in the LIGO Observatories to be used in the near future.

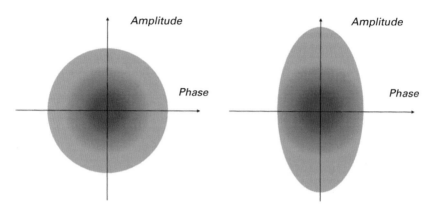

Figure 9.7
Left: The plot shows the uncertainty for the amplitude and for the phase when no gravitational wave is detected (since it is at the origin, the average amplitude is zero). Right: The plot shows that the uncertainty in phase (which is what is measured) is squeezed at the cost of increasing the uncertainty in the amplitude (and the radiation pressure noise).

9.7   Optical Tricks and the Need for Control

If the detector arms were longer than 4 km, the detector would be more sensitive. But if it is too long, it would lose sensitivity as well, at least at some frequencies. For example, let us consider a gravitational wave of 100 Hz. Such a wave would change the distance between the mirrors with a period of 10 milliseconds. The laser beam travels the 4 km at the speed of light and reaches the end mirror in only millionths of second. During that time, the distance has changed by just a small fraction (0.1 percent) of the maximum amplitude of the oscillation. If we make the arms longer, the measured fraction would be larger. But if we make them too long, the beams take too long to travel along the arms, and they would have begun shrinking again, so we measure a smaller fraction again. A 100 Hz gravitational wave traveling at the speed of light has a wavelength of 3,000 km. In addition to being impractical on Earth, an interferometer that long would be more expensive. The length chosen was a compromise between cost and performance.

To achieve a similar goal, something else can be done: make the laser beam go back and forth many times, so it will accumulate the fractions of the wave that it measures each time, effectively increasing the signal. This effect is achieved by making the laser beam bounce back and forth several times between the mirror at one of the ends of the interferometer masses and an additional mirror close to the beam splitter. There are two ways of doing this: one design uses "delay lines," and the other uses "optical cavities."

The interferometer prototypes of the 1980s used a delay line, where the (green) laser beam was bounced off different points of the mirror's surface. Such a technique requires large mirrors, and fractions of the light beams are often scattered from one beam to another when reflected at the mirrors, producing noise. In present-day detectors, the laser beams are sent back and forth by bouncing them at the center of the mirrors, with the beams interfering constructively with each other (see figure 8.3) in an optical resonant cavity. This is known as a Fabry–Pérot cavity in honor of Charles Fabry and Alfred Pérot, French physicists who developed it in 1899. It is a very common technique in optics: the laser itself is a cavity of this kind. To increase the number of trips that the laser beam makes, the mirror close to the beam splitter (the "input" test mass, or ITM) is not completely reflective. The mirror at the end of the interferometer arm (the "end" test mass, or ETM) is as reflective as the technology allows. The net result is that a small fraction of the beam (about 30 percent) is transmitted first by the ITM, but it returns, transmitting about 30 percent of the light to the detector output with the rest going back to the ETM, and then repeating the cycle. These fractions add up, effectively increasing the interferometer's length by a few hundred times. The transmissibility of the ITM determines the equivalent number of times that the light propagates along the arm and is therefore the effective length.

This technique has a complication: the beams in the cavity have to be in phase with each other (the crests of the light waves have to coincide). To achieve this, the cavity created between the two mirrors must have the correct length. When it does, the cavity is said to be "in resonance," a very common term in physics. It is also necessary for the mirrors to be very well aligned, with the centers of their surfaces perpendicular to the laser beam. If the laser beam does not bounce perpendicularly, it does not go back on itself. This alignment is achieved by pushing and pulling the mirrors so that they are at the correct distance and orientation. In Initial LIGO, four magnets were glued to the back of the mirrors, and four electromagnets were put on the frame from which each mirror was hung. These electromagnets pushed or pulled the magnets as needed to place the mirrors to be in their correct positions. To know how much to move the mirror, a signal with the information about the difference in phase of the light beams going back and forth is needed.

One method for sensing the distance between the mirrors in a cavity is called the "Pound-Drever-Hall" method, after Robert Pound of Harvard University; Ronald Drever of the University of Glasgow and later Caltech; and John Hall of the Joint Institute for Laboratory Astrophysics in Boulder, Colorado (Hall received the 2005 Nobel Prize for his work on optical combs). This method is used nowadays in many other systems. It uses similar techniques to those used in radio communications, such as frequency modulation (modulating in this context means "changing"). The phase of the laser is modulated (like radio waves emitted by an FM radio station; FM stands for "frequency modulation"), and the light reflected by the cavity (coming back from the input test mass) is demodulated (which is what an FM radio receiver does). Despite its complexity, it is an extremely valuable method, especially when using modern digital systems. Electronic devices convert the photocell signal into numbers, which are sent to a computer that processes them. The result is converted into electrical currents sent to the electromagnets. The whole process is fully automatic.

When the cavities are in resonance, the laser beam will bring information about the length of the arm as if it were much longer: the detector is known as a Fabry–Pérot–Michelson interferometer. As mentioned, if the lengths are equal, the interference of the light beams form both arms is destructive: this is called "operating at the dark port." In this case, the detector operates like a mirror. This yields another advantage: the returning light can be recycled by placing a partially transmitting mirror between the laser and the beam splitter. Many more photons are thus circulating in the system, which, as already discussed, effectively reduces the quantum noise. This requires fine tuning the distance between the beam splitter and the new recycling mirror. This technique is known as "power recycling." So the LIGO interferometers are properly called "power recycled Fabry–Pérot–Michelson interferometers." This configuration was used by Initial LIGO.

Few changes were made to this configuration for Advanced LIGO, but one was particularly important: another recycling mirror was added, this time at the output. This configuration is known as "signal recycling," although there is no recycling of the gravitational wave signal but an increase in the sensitivity at certain frequencies. This is known as a "dual recycled Fabry–Pérot–Michelson interferometer." Plate 4 shows a schematic diagram of the laser beams and the mirrors used.

This configuration required low-noise electronic circuits and a sophisticated computer-controlled digital system to keep all optical cavities resonant. Without these feedback-control systems, even if all masses are at the right places with the correct alignment, they will drift in fractions of a second, and there would no signal to read at the output. Unfortunately, these systems also add noise as well. The control systems are complicated enough that just making the detector achieve an operational state in which the noise can be measured is a cause for celebration itself, as shown in figure 9.8. Only

Figure 9.8
Scientists in the LIGO Hanford Observatory control room, celebrating on December 12, 2014, the first time that an Advanced LIGO detector achieved an operational state (i.e., was "locked"). (Credit: Caltech/MIT/LIGO Lab.)

then can the time of commissioning start, when many different—and sometimes unexpected—sources of noise are discovered. But in the process, so are different ways of improving the sensitivity, including techniques that were not part of the original design but proved to be effective and ready to be used.

9.8   Thermal Noise: If It's Warm, It Moves

We are not yet done with discussing the different sources of noise that can affect a clear detection of the signal. We are missing one of the hardest noises to reduce: the "thermal noise." This noise has its origin in the microscopic nature of all matter: everything is composed of atoms that are constantly moving. The higher the temperature is, the faster the vibration of the individual atoms becomes, and therefore larger the thermal noise. This microscopic atomic motion translates into motions of the macroscopic quantities of the system, for instance, the position of a test mass in LIGO. These types of motions driven by fluctuations are also known as Brownian motion, because these motions were discovered by the Scottish botanist Robert Brown in 1827, when the modern concept of atom was not known. He studied pollen diluted in water through a microscope. He saw that the pollen particles moved randomly instead of in straight lines. It was as if these particles were alive, which Brown disproved by observing the same effect in inorganic matter. It was Einstein in 1905 who wrote an article proposing that the effect was because water is composed of molecules and the particles of pollen moved due to collisions with the water molecules. This was one of the four articles published in Einstein's annus mirabilis (1905).

The total amount of thermal noise depends only on the macroscopic features of the system (mass, size, etc.) and the temperature. However, we are interested in how much of that noise exists in the sensitive frequency band of the detector, where we want to find gravitational waves. There is a theorem in thermodynamics that relates the level of thermal noise at different frequencies with the way the system loses energy when it moves and its resonant frequencies. A system made of two masses connected by a spring has a well-defined resonant frequency that depends on the values of both the stiffness of the spring and its mass: when energy is input to this system at this right frequency, it will oscillate with maximum amplitude. If there are more masses connected by the spring, there will be several resonant frequencies (depending on the values of the masses). The LIGO mirrors are suspended, and there are "pendulum" frequencies for motion along the six degrees of freedom of the masses, three displacements (up/down, back/forth, left/right), and three possible rotations, (pitch, roll, and yaw, as for aircraft). The suspension fibers, like the strings of a violin, also have resonant frequencies (like the notes of a violin) that also influence the motion of the suspended mass. An elastic body—the LIGO mirrors themselves—can be thought of as many masses connected by springs. The LIGO test masses have an infinite number of resonant frequencies, but the ones that most affect the detector noise are those that move the mirror in the horizontal direction:

the pendulum mode, the "violin modes" of the suspension fibers, and the "drum modes" of the mirrors (their deformations are akin to those of a drumhead).

The abovementioned theorem in thermodynamics (known as the fluctuation-dissipation theorem) essentially states that the thermal noise in any system is mostly concentrated at the resonant frequencies of the system. The degree of concentration of this thermal noise depends on how the physical system loses energy, which depends on its constituent materials. The results of this theorem are the leading criteria for choosing the materials for the mirrors and other components. These materials should be such that the system will dissipate as little energy as possible. That makes the noises mostly concentrated at the resonance frequencies, and therefore there is less noise at the frequencies of interest for detecting gravitational waves. For Initial LIGO, the mirrors were made of fused silica (which has little energy loss), and they were hung using steel fibers. Magnets were also glued to the mirrors, which were a further source of loss of energy. In Advanced LIGO, the mirrors are still made of fused silica, but they are suspended from fused silica fibers (they lose less energy than does steel) and are fused monolithically to the mirror of the same material, so there is no friction or loss of energy at that point (this technique was developed for LIGO mirrors at the University of Glasgow laboratory).

The mirror still needs to be moved to keep the cavity in resonance. Since a quadruple pendulum is used in Advanced LIGO, the magnets are glued to the mass from which the final mirror is suspended. The electromagnets are in another mass hanging from a quadruple pendulum behind the principal mass. This configuration does not supply enough force when the resonance in the cavity needs to be sustained: in the bottom pendulum shown in the right panel of figure 9.5, the reaction mass in the back has gold rings, where high voltages are applied to push on the test mass. Once again, we can see that the complexity has increased considerably: it is the price paid to improve the sensitivity by an order of magnitude.

9.9 Summing Up

When the detector is in operation, even when no gravitational wave is detected, noise is always present. Gravitational wave scientists characterize it with a plot known as a noise curve. It shows the amplitude of the noise at different frequencies (see figure 9.9). From estimates of the noise due to the different components installed in the detector, there is an expectation for the noise curve: this is often called the "design sensitivity curve." Ideally, after all the components of the detector are installed, the noise should be equal to this design sensitiviy curve. In reality, it never works like that, particularly when the instruments are so complex. There are always sources of noise produced by the thousands of control systems and the environment that need to be measured, diagnosed, and eliminated, one by one. It takes years to reach the optimal sensitivity that was originally planned. During this process, a better understanding of the instrument emerges, and sometimes new ways of increasing the sensitivity beyond the design are discovered, enhancing the search for gravitational wave sources.

Which source of noise is largest in the design sensitivity depends on the frequency being considered. For the most sensitive range of frequencies (the one for which gravitational waves would clearly stand out, around 100 Hz), the dominant noise is the thermal one. At lower frequencies, seismic noise is more prevalent; and at higher values, the predominant source is quantum noise. As the figure shows, all these sources were significantly reduced in the design of Advanced LIGO by using more laser power, replacing simple pendulums by quadruple ones, limiting energy dissipation, changing

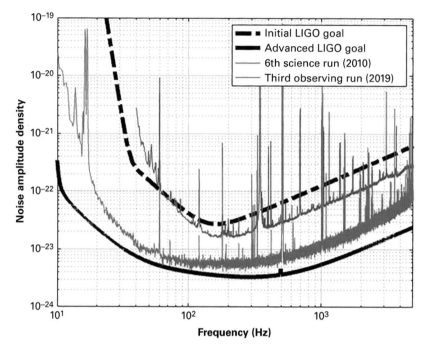

Figure 9.9

Noise amplitude at different frequencies: the lower the noise, the more sensitive the detector is to gravitational waves. The continuous black curve is the design sensitivity curve for Advanced LIGO. The dashed black curve was the goal for Initial LIGO, which was achieved in 2005 and exceeded in 2010, like the curve measured during S6 shows. In 2019, the noise amplitude in the Advanced LIGO curve is not as low as the goal, but there has been progress since the beginning of 2015.

passive by active seismic isolation, and improving in general the control systems for the positioning of different subsystems.

Figure 9.9 shows the noise curves that were the goal of Initial LIGO and Advanced LIGO. When the noise present in the instrument is measured—like the curves shown in the figure—not only is the amplitude bigger than the goal, there also appear many vertical lines with much larger amplitudes. These lines are due to sources that contribute to the noise at a given frequency, like the lines that are multiples of 60 Hz (60, 120, 180, . . .). These have their origin in the frequency of the alternating current provided by the electrical power grids in the US (60 Hz). Other lines of large amplitude are the "violin modes" (the natural resonant frequencies) of the fibers that suspend the mirrors. Yet others are the result of sinusoidal forces used to calibrate the instrument.

When a noise curve is shown as in 9.9, it is implicitly assumed that the noise is a random signal, where its amplitude at each frequency is described by the curve. However, this is not true—there are sometimes very large transients ("glitches") at the output of the LIGO detectors, which scientists know are not gravitational waves (because there is no similar signal in the other LIGO detector). These transients contribute to noise in a different way: they obscure the smaller gravitational waves. Identifying the origin of lines in the noise spectrum, and of glitches, keep busy the "noise detectives" in the Collaboration who try to eliminate them. They are a true army of many dedicated and hardworking scientists, who are so essential that they form an official group in the collaboration with a special name: the Detector Characterization Group.

The amplitude of a gravitational wave can be compared with the amplitude of the noise at each frequency and summed over all frequencies. This leads to an important quantity known as the signal-to-noise ratio. This quantity will depend on the details of the source and the distance to it. Evaluating it for a given source allows researchers to summarize in a simple way how sensitive the detector is. By convention, the signal-to-noise ratio is evaluated for a collision of two neutron stars. The latter have a rather narrow range of masses, and therefore both can be considered "identical" for this purpose. LIGO chooses neutron stars of 1.4 solar masses, which is typical. The choice of other sources, like binary black holes, is not such a clear universal standard. The lower the noise level, the farther the distance at which neutron stars are detected. This maximum distance is known as the binary neutron star range of the interferometer.

When Initial LIGO data started being analyzed in 2002, the instruments were sensitive only to collisions of neutron stars in our galaxy (less than 100,000 light-years). Such events should occur once every 30,000 years. By 2010, the range had increased to 60 million light-years, enough to see collisions in the Virgo cluster of galaxies. But even at this sensitivity, a merger of neutron stars is expected to happen only every 50 years. The Advanced LIGO technology was necessary to reach farther away (eventually to 500 million light-years) and detect many gravitational waves per year. The signals produced by black hole collisions have a larger amplitude (black holes are more massive than neutron stars), and the detectors can see them farther away. The first detection of gravitational waves, on September 14, 2015, was a collision of two black holes more than 1,000 million light-years from Earth. The first detection of colliding neutron stars took place on August 17, 2017, at 130 million light-years from Earth.

At the time this book is being written, in 2022, Advanced LIGO has not yet achieved design sensitivity, as can be seen in figure 9.9. But the noise has been reduced enough to see binary neutron stars mergers 430 million light-years away. There are other detectors in the world, and there are ideas and plans to improve further the LIGO observatories' sensitivities. Concepts and plans for future generations of detectors 10 times more sensitive are also in the making. Just as centuries after Galileo's first telescope optical instruments are still being built and improved, the LIGO observatories represent just the beginning of a new astronomy that is expected to evolve and improve quickly in the future.

10
At Last: Detections—and Many!

The Initial LIGO interferometers started scanning the universe in 2002. Data were acquired and analyzed for the first time that year. Improvements based on better understanding the sources of noise (and reducing it) were implemented progressively during subsequent years. The sought-after sensitivity (called "design sensitivity") was achieved several years later, in 2005. Data were taken and analyzed in 2005–2007, and although no detection was made with Initial LIGO, the runs showed that the laser interferometric technique was capable of achieving the great precision it had been designed for. No scientist thought that there was a high chance of observing a gravitational wave, but the successful proof of concept shown by kilometer-scale instruments achieving their planned design sensitivity surprised many and raised the expectations for a new astronomy.

Some of the conclusions from *not* having observed gravitational waves operating at such sensitivity allowed important astrophysical conclusions. Among them was obtaining a new experimental limit on the amplitude of gravitational waves from the early universe (at the frequencies within the LIGO sensitivity range), which was better than the limits provided by earlier experiments. Furthermore, an agreement was reached between the LIGO Scientific Collaboration and the Virgo Collaboration to combine the data from the LIGO and Virgo instruments. It was also decided that all papers to be published would be signed jointly by the scientists from both collaborations.

The success in achieving the designed sensitivity offered a well-defined roadmap to increase that sensitivity with advanced detectors and reach the threshold of detection for the first time. The National Science Foundation approved the investment in the advanced technologies funding Advanced LIGO; the European funding agencies did the same supporting the development of the Advanced Virgo detector.

The stages of Initial LIGO detectors represented many years of incredible efforts by hundreds of scientists and engineers. The successful operation prepared a whole generation to tackle the formidable challenge of a first detection. The end of Initial LIGO was itself a scientific landmark.

Five years of hard work started in 2010 to install all the new technologies that Advanced LIGO required. By 2015, more than 20 years after the construction of the observatories, Advanced LIGO was ready. The advanced detectors had not yet reached their design sensitivity, but were about three times more sensitive than Initial LIGO, so a data-taking campaign of a few months had been planned to start in September 2015. While getting ready for this first observing run (O1),[1] a gravitational wave was unexpectedly detected, throwing LIGO scientists right into the maelstrom of a new astronomy. It was like the cast was on stage ready to start the play, but the curtains went up a few minutes earlier than anticipated.

In previous chapters, we have summarized the history of relativity, the theory, the methods to model the waveforms, and the instrumental challenges. In this chapter, we describe the exciting first steps of gravitational wave astronomy with Advanced LIGO.

10.1 The First Detection: GW150914

On September 14, 2015, at 2:50 a.m. in Hanford, WA, and at 4:50 a.m. in Livingston, LA, the interferometers at both LIGO observatories were operating normally. The official first science observational run, called now "O1," was scheduled to start in a few days. The instruments were operated through what is called an "engineering run." The detectors were undergoing testing and calibration in preparation for O1, which was set to start on September 18. During an engineering run, data are still obtained, and the analysis algorithms are tested. Several instrumental diagnostics of different types are also made. A system used to inject simulated signals and an automatic alert system that listed candidate events on web pages accessible to the Collaboration researchers were also being tried for the first time with the new instruments.

The operators on duty at the sites were the physicists Nutsinee Kijbunchoo in Hanford and William Parker in Livingston. Kijbunchoo was a graduate of Louisiana State and Parker of Southern University, a historically Black college in Baton Rouge, Louisiana. Researchers in Hanover, Germany, where the time was 11:50 a.m., and early-riser researchers in Florida (5:50 a.m.) noted that the web pages created by the analysis codes were indicating a strong candidate.

But this candidate, lasting only a fraction of a second, had an amplitude larger than the typical false positives. It was also larger than the expected amplitude—at that time—for gravitational waves, even for black holes. However, a simple plotting of the data, known as a time-frequency representation, showed a textbook plot for the characteristic waveform of a collision of two black holes (see figure 10.1). This surprised researchers, who immediately contacted the observatories to see whether simulations of signals (injections) were being carried out. Kijbunchoo, in Hanford, had briefly stepped out of the control room, so no one answered the phone. But in Louisiana, Parker did pick up and confirmed that the interferometers

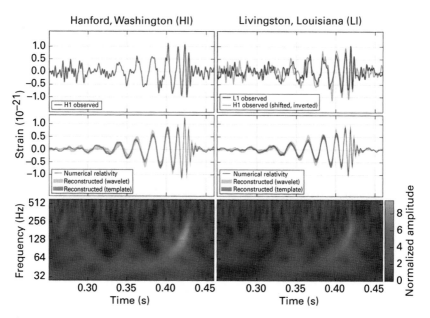

Figure 10.1

The gravitational wave event GW150914 observed by the detectors at LIGO Hanford (H1, left column panels) and at Livingston (L1, right column panels). Times are shown relative to September 14, 2015, at 09:50:45 Coordinated Universal Time. Top row, left: H1 strain. Top row, right: L1 strain. GW150914 arrived first at L1 and 7 ms later at H1; for a visual comparison, the H1 data are also shown top right, shifted in time by this amount and inverted (to account for the detectors' relative orientations). Second row: Solid lines show a numerical relativity waveform for a collision of black holes of 29 and 36 solar masses. Shaded areas show two independent waveform reconstructions from the data. One (dark gray) models the signal using binary black-hole template waveforms. The other (light gray) does not use an astrophysical model. Bottom row: A time-frequency representation of the strain data, showing the signal frequency increasing over time. This is a plot showing how the frequency (values going up on the vertical axis) varies with time (increasing to the right on the horizontal axis). The shading represents amplitude, with the lighter shading representing an increased value. (Credit: Adapted from *Physical Review Letters* 116, 061102, 2016.)

were working normally and no tests were being run. Remarkably, two physicists had been carrying out tests in Livingston, studying the influence of a car passing close to the concrete tubes enclosing the interferometer. They were using a GPS unit that ran out of battery power 50 minutes before the event and caused them to suspend the tests. The signal could have been missed if these tests had continued.

Depending on the location of the source in the sky, a signal could arrive at one observatory before the other, or simultaneously at both (if it came from a source equally distant from both). The time taken for light (or gravitational waves) to go straight from one observatory to the other is 10 msec. Clearly, the arrival times difference cannot be longer than that. Livingston received the signal 7 msec before Hanford did, so it was compatible with a signal of astrophysical origin.

But someone may ask: given that the signal was so intense, why were signals of the same type not detected at Initial LIGO? A more careful analysis shows that in spite of its amplitude, given the higher noise in Initial LIGO at the frequencies of this wave (near the lowest frequencies of the sensitive band), a signal like this would not have been detected.

At least inside the Collaboration, the genie was out of the bottle: this could be the first detection. The authorities of the LIGO Lab and the LSC Collaboration at the time (David Reitze and Gabriela González, respectively) reassured members of the Collaboration that it was not a blind injection, a technique that had been used in the past to test analysis algorithms. But many did not believe them. After all, that is what they were expected to say precisely if a blind testing signal had been injected into the detectors. Rumors immediately started to circulate among the international scientific community, and science journalists started asking questions. These were stressful days for David (Dave) and Gabriela (Gaby), who had to quell speculations within the international scientific community. They were in a complicated position: outside the collaboration, people did not understand why such a standout signal could not immediately be announced as a discovery.

But inside the collaboration, it was well accepted that a rigorous analysis of the event was required before the signal could be accepted as real. A careful vetting would avoid the possible embarrassment that an early announcement had to be followed by a later retraction. Advanced LIGO had barely started taking data for a few days at the time of the detection. Analyzing it through some reasonably extended period of time was needed to characterize the instrumental noise. The estimation of the event happening by chance, and not coming from an astrophysical source, could be properly assessed only after the detector behavior was properly understood. The original intention was to make a 3-month-long observation run. Under the pressure to analyze the first candidate, the collaboration decided to carry out the analysis with just 2 weeks of data. But interferometers do not operate 24/7 uninterrupted. They can go offline for several reasons: earthquakes or tremors from the ground, thunderstorms, or unavoidable maintenance could force operations to come to a halt. The instruments had to operate for 5 weeks for the collaboration to obtain 2 weeks' worth of data with both observatories working at the same time. The statistical analysis of the data stream took about another month. Eventually, the conclusion was clear: this signal could have happened by chance only once in 203,000 years. The unveiling of this result was done in a virtual meeting of the collaboration where there was great suspense and anxiety, and it generated applause and cheers: gravitational waves had been detected! Moreover, they had been produced by a collision of two black holes located 1,300 million light-years away from the Earth. This result meant that the collision happened 1,300 million years ago! That was when multicellular organisms were appearing on Earth for the first time. They would surely not have imagined that their descendants were going to build devices to detect that wave.

The masses of the black holes also generated some surprise and incredulity: one was 29 solar masses and the other was 36 solar masses. The final black hole was 62 solar masses, 3 solar masses less than the sum of the masses of the colliding black holes. It turns out that this missing mass had been radiated as energy through the emitted gravitational waves (remember Einstein's formula $E = mc^2$). The intensity (quantity of energy per unit of time) during that emission exceeded the intensity of the visible light of all the stars in the universe combined!

This was all very surprising: none of the stellar-mass black holes identified by astronomers at the time (all observed indirectly through the electromagnetic emission from accretion disks swirling around them) were heavier than 20 solar masses. If confirmed, this result was not only going to be the first observation of gravitational waves, but possibly the first glimpse into a new "species" of black holes. Since even at the time of the writing of this book, there is no consensus about how black holes of such large masses are formed, this added pressure on the members of the Collaboration to be sure that their analysis was correct. The codes for computing the parameters of the system and the corresponding uncertainties take a significant amount of time to run after a detection is made. Furthermore, the codes required had not yet been thoroughly reviewed. Could this be a joke by people trying to fool the collaboration members into believing that a discovery had been made, followed by humiliation? There was intense scrutiny of this possibility, in a study led by Matthew Evans, a faculty member at MIT, looking for consistency of the signal in the many data streams that are used in the detectors. But nothing had been added to the recorded data. Perhaps somebody had somehow pushed the mirrors at both detectors with something else, simulating the effect of a gravitational wave? The conclusion was if this was a hacking job, it would have taken more than one person with deep and current knowledge of the detectors, and with almost superhuman characteristics to be able to carry out this hack and leave no traces. The many detections that have followed the first one (89 more in 2021!) alleviated any doubts—but those were intense times.

A paper was prepared to submit for publication, and many other papers with details were being written as well—many researchers were very busy. Although the detection had been made by the LIGO instruments, the analysis had been carried out by groups of the LIGO and Virgo collaborations together, so it had been decided well in advance that the papers would be signed by both collaborations. The collaboration leaders, Fulvio Ricci and Gaby for Virgo and LIGO, respectively, appointed a paper-writing team, led by Peter Fritschel of MIT and Pia Astone from the National Institute of Nuclear Physics in Rome. The team was composed of six scientists who had to take into account comments about the drafts of the manuscripts by all members of the collaborations. It was an intense task that generated many drafts, each circulated to the collaborations to gather comments (thousands of them). The final version had to be voted on by both collaborations. The discussion by the LIGO Collaboration took place in a virtual meeting in mid-January 2016, and more than 300 people participated in it. There was a long discussion about what details to include in the final paper. It seemed like the meeting was never going to end. Finally, Gaby closed the debate and a vote took place. It was almost unanimous, and the meeting ended with everyone applauding.

It had been previously decided—incredibly, shortly before September 14—to make the public announcement when the paper was accepted for publication in a prestigious scientific journal after undergoing rigorous peer review. In case of acceptance, the article with the details would still be kept secret until a public announcement in a press conference could be scheduled. The article was submitted to the journal *Physical Review Letters*, published by the American Physical Society (APS). The journal promised that the article would be evaluated as fast as possible. By the end of January 2016, it was accepted for publication.

That day the secret of the detection was almost compromised beyond control. The journal website publishes on a daily basis an electronically generated list of accepted papers. Somebody noted in Europe that the website was listing the LIGO–Virgo paper! The Virgo spokesperson at the time, Fulvio Ricci, telephoned Gaby. It was 3:30 a.m. in Louisiana. Fortunately, at that time, Jorge—Gaby's husband—was the lead editor (and founder) of another APS journal, *Physical Review X*, and he had at hand the cell phone number of *Physical Review Letters* editor, Robert Garisto. Initially, Garisto did not pick up the phone. After repeated calls, he realized they were coming from Louisiana, and he decided to call back. The article was removed from the web page in less than half an hour.

The LIGO collaboration and the LIGO Laboratory directors jointly with the Virgo collaboration agreed that NSF would organize the announcement in the US, in parallel with an announcement in Italy. The NSF was scoring a big win after having bet on the most expensive scientific project in its history. NSF organized the press conference in Washington, DC, on February 11, 2016, at the National Press Club. Founded in 1908, the National Press Club has been used for announcements by many US presidents, starting with Theodore Roosevelt, as well as by many heads of state. Rumors became more intense as the date was approaching. On February 10, the people in charge of making the announcement—Reitze, González, Thorne, and Weiss—had a rehearsal at the NSF headquarters, at the time in Arlington, Virginia, a suburb of Washington, DC. That evening they went for dinner at a restaurant near the building and happened by chance to meet a group of well-known relativists who were at NSF in a panel evaluating grant proposals at the same time. The irony is that both the presence of those panelists as well as that of the people making the announcement was confidential. One of the panelists told the LIGO members: "You are not here, are you?" and Reitze responded: "No, and neither are you, right?"

The hall of the National Press Club is very august, covered in oak panels, but is quite small, particularly when many TV cameras are present, as they were on February 11. A photograph of the press conference is shown in figure 10.2. An annex had to be prepared,

Figure 10.2

Press conference held at the National Press Club on February 16, 2016. From left to right: France Córdova, then director of the NSF, presiding at the press conference; David Reitze, executive director of the LIGO Laboratory (Caltech/MIT); Gabriela González (Louisiana State University), then spokesperson of the LIGO Scientific Collaboration; Rai Weiss (MIT) and Kip Thorne (Caltech), pioneers of the LIGO project and awardees of the Nobel Prize in 2017. (Credit: NSF, clip from recorded video.)

where some journalists and members of the Collaboration watched the announcement by closed-circuit TV. The announcement was widely covered in the media, appearing on the front page of the *New York Times* and newspapers around the world. President Obama tweeted about it. In a Baton Rouge newspaper, it occupied the upper part of the front page above the news that then-presidential candidate Donald Trump had been in town.

That night, the members of the LIGO collaboration present in Washington organized a celebration in a bar; on one of the TV screens, the Syracuse University basketball team was playing a game. At halftime, a program on the bar TV showed how Peter Saulson, former LSC spokesperson (and Gaby's PhD thesis advisor) and his LIGO group at Syracuse were honored at the court. This is the highest form of recognition in the eye of the public at many universities! The LIGO members put up a banner at the bar celebrating the discovery. At closing time, a group that did not belong to LIGO was sitting beneath it. When their permission was asked to remove the banner, this random group of people asked: "Are you the physicists from Louisiana that are in the *New York Times*?"

By sheer chance February 11, 2016, was the first time that the International Day of Women and Girls in Science, declared by the UN General Assembly was celebrated.[2] Scientists in Spain were trying to make the topic trend in Twitter, something they managed to do until they were displaced by the LIGO announcement. It was nice that on this day, two of the five people making the announcement at the National Press Club, Gaby and France Córdova, were female.

Shortly thereafter, González, Reitze, David Shoemaker (who would succeed Gaby in 2017 as spokesperson) representing LIGO and Fleming Crim, head of the NSF Directorate for Mathematical and Physical Sciences, were invited to testify about the discovery at the Committee of Space, Science and Technology of the US Congress. This was the same committee in front of which Tyson had testified in 1991 against constructing LIGO. In the question-and-answer period, it was clear that there was now bipartisan enthusiasm from lawmakers about the discovery and the LIGO effort.

10.2   The Second Detection: GW151226

While the first month of O1 data was analyzed to confirm the first detection of gravitational waves, LIGO continued operating in observing mode. The original plan (before having any indication of signals) had been to take data for 3 months and conclude before the December holidays of 2015. However, the first detection was so extraordinary that there was great interest in trying to detect a second one. It would not change the statistical value of the first result, but it would have a great psychological impact, apart from being another scientific discovery. There have been some physics discoveries with high statistical significance that were not observed again, and results of this kind open the doors to reasonable skepticism. A second candidate signal was found on October 12—also a collision of black holes—in the analysis done for the first detection, but it was too weak to warrant an observational claim. Because of this, it was decided that the run would be extended until mid-January 2016. This took a great deal of leadership by the Hanford and Livingston Observatory heads, Fred Raab and Joe Giame, respectively, who had to ask operators, technicians, engineers, and scientists at the sites to keep working, after very intense months, at a time when many had planned to go on vacation. But the sacrifice paid off: on December 26 (Coordinated Universal Time, still Christmas in the US), another collision of smaller black holes was detected. Their masses were 14 and 7.5 solar masses. The amplitude, smaller in spite of the collision happening at about the same distance from Earth as the first collision detected, was nonetheless commensurate with the smaller masses. The lower masses also implied a higher frequency. But this fact allowed the interferometers to capture the signal for a longer time than the first gravitational wave detected (which lasted a mere fraction of a second). This signal was observed for many more cycles before the merger than the first detection. The lower amplitude but longer capture compensated for each other in terms of detectability. The statistical significance of the resulting analysis confirmed a detection with comparable certainty to the first one. By the time of the announcement, there had not been a complete analysis of the second detection, so the second detection wasn't disclosed at the press conference. But now LIGO and Virgo scientists were more confident that they had entered the new era of black-hole gravitational wave astronomy.

10.3   More Detections: Catalogs of Gravitational Waves

The O1 campaign concluded with outstanding success on January 19, 2016. The sensitivity of the two LIGO instruments had been improved, and the second observational campaign (O2) started on November 30, 2016. It ended 9 months later, on August 25, 2017.

As we discussed, an indicator of the detectors' sensitivity is the farthest distance at which a neutron star collision could be detected. This value increased from 200 million light-years in O1 to 260 million light-years at the beginning of O2, and to more than 320 at the end of the second campaign. For comparison, Initial LIGO had a considerably shorter range: 60 million light-years. The detected black holes were at much farther distances due to their much larger masses compared to neutron stars.

The combined duration of O1 and O2 was 14 months, or slightly longer than a year. The times analyzed were 170 days when the detectors were operating in coincidence, about 40% of the time. But O2 involved a significant development: on August 1, 2017, Advanced Virgo, the European sister of Advanced LIGO, joined the search. Near the end of O2, there were now three interferometers scanning the sky. As we will see later in this chapter, this newly formed hunting party made a terrific catch within days of starting joint operations.

By the time O2 ended, 11 events had been identified among the candidates as having probabilities higher than 50 percent of being of astrophysical origin. The results were compiled into the first catalog of gravitational wave sources, made public in December 2018 with the name GWTC-1 (Gravitational Wave Transient Catalog). The detections in this catalog are shown in figure 10.3.

Before gravitational waves from black-hole mergers were detected, knowledge about the masses of stellar black holes was obtained through X-ray observations obtained over the past couple of decades.

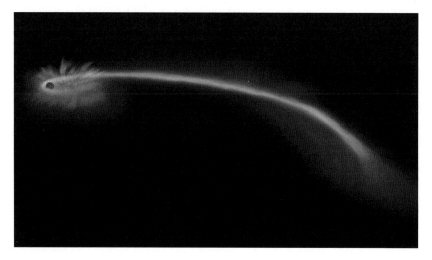

Figure 10.3
Tidal disruption of a star by a black hole. The artistic rendition shows a disk of stellar debris around the black hole in the upper left of the illustration, and a long tail of debris that has been flung away from the black hole. This tidal disruption was dubbed ASASSN-14li and was discovered in an optical search by the All-Sky Automated Survey for Supernovae (ASASSN) and followed closely by X-ray telescopes from the NASA's Chandra X-ray Observatory, the Swift Gamma Ray Burst Explorer, and the European Space Agency's XMM-Newton. (Credits: artist conception based on image by NASA/ CXC/U. Michigan/J. Miller et al.; illustration by NASA/CXC/M. Weiss.)

There are about a couple dozen of them in X-ray binary systems, mostly in our galaxy, with the heaviest black hole having about 20 solar masses, and the rest between 5 and 16 solar masses. The masses of the 20 black holes before merger in the first catalog were between 7 and 50 solar masses. The 10 final black holes resulting from the mergers are of course heavier, between 18 and 80 solar masses. All events happened very far from our galaxy, at distances ranging from about 1 to 9 billion light-years away! This was truly a new way to do astronomy of the darkest and most mysterious objects in the universe.

There was a very special event, GW170817, that merits (and has) its own chapter (chapter 11) in this book: the merger of two neutron stars, observed not only with gravitational wave detectors but also with ground and space telescopes of many different wavelengths, from radio to X-rays, including optical signals and gamma ray bursts. But before describing that spectacular discovery, we have more discoveries in new catalogs.

10.4   Even More Detections: New Catalogs

In April 2019, a 1-year run with LIGO and Virgo was started, known as O3; it paused in October to improve the sensitivity of the detectors and resumed in November 2019 with the intention of taking data until the end of April 2020 or possibly longer if the Kamioka Gravitational Wave Detector (KAGRA; see chapter 12) joined the observing run. Unfortunately, the observing run was cut a month short on March 27, 2020, by the Covid-19 pandemic.

Freed from the worries of releasing false alarms, the LIGO-Virgo collaboration decided that for O3, the announcement of possible candidates was going to be public and immediate. There were 56 alerts (and a few retractions) from different kinds of candidate events, including neutron stars and black holes: this was more than five times as many events as in the first catalog! The average rate of alerts was more than one per week—although none yielded observations of electromagnetic counterparts as did GW170817, possibly because the events were mostly black-hole mergers and were far away.

The analysis of the data was undertaken even during the pandemic, and two new catalogs were published with the results: GWTC-2 with the analysis of the first 6 months in the first part of the run (O3a), and GWTC-3, with the last 5 months of data (O3b). In GWTC2.1 (a deeper analysis of the data after GWTC-2) there were 44 events, and in GWTC3, there were 35 more events. Including the 11 events in GWTC-1, there are now 90 observations of gravitational waves produced by the merger of compact objects, in just 6 years since September 2015—a new era for astronomy!

Most of the new O3 detections are mergers of black holes, just as in O1 and O2. But several different and very interesting events were detected: another merger of neutron stars (but four times farther away than the first one, which is described in chapter 11); two mergers of a neutron star and a black hole (never observed before); and mergers of very heavy black holes (with the largest couple having 65 and 95 solar masses).

Moreover, there are now observations of several compact objects that are heavier than neutron stars can be (the upper limit is about 2.5 solar masses) and lighter than known X-ray black holes (about 4–5 solar masses). Such mass values had been hypothesized to be in a "mass gap" in compact objects separating neutron stars and black holes. From gravitational wave observations, the objects resulting from the merger of the neutron stars fall in this mass gap; there are also two observations of mergers of a black hole with an object in this gap.

As shown in plate 6, this collection of discoveries significantly increased our knowledge of the population of black holes and neutron stars as a whole. It also allowed us to appreciate the relation between the final masses of the compact objects and their progenitors and to make a comparison with the already known black holes in our galaxy. In only 5 years, we are already doing black-hole astronomy with many systems available to study common patterns, and many more to come. A new era indeed!

10.5   The Properties of the Detected Sources

The characteristics of each detected system (pairs of black holes or neutron stars in the last moments of their dance before the final collision) can be described with 15 numbers. These represent physical quantities, "parameters" in physics jargon. Eight of them describe intrinsic properties of the binary pair, like their masses and their spins before and after merger. The rest of the numbers describe the position of the system in the sky, its distance from the Earth, the inclination of the orbital plane of the system, and so forth. A sophisticated and time-consuming analysis using different waveform models is needed to estimate those parameters. Another number (the first one we estimate) is the time of collision, of course.

In the case of a system composed of neutron stars, there exists an additional parameter of interest describing how the shapes of the stars deform. The deformation is due to gravitational attraction and is of the same nature as the deformation induced in the Earth's oceans by the presence of the Moon: the tides. Depending on the relation between the masses of the stars, they can experience deformations so large in their structure that in the process they lose enormous quantities of matter and can even be destroyed. Binary black holes also deform each other, but there is no matter involved, only space-time. Moreover, Einstein's general relativity prevents the destruction of black holes before they merge. The observation of the deformation of neutron stars can provide clues about their structure. In intuitive terms, a "harder" star would deform less than a "softer" one. The internal energies and pressures involved in these astrophysical objects could never be explored through experiments on Earth. Even though it is clear that neutron stars exist—we have observed more than a thousand of them—what is known about their internal structure is considerably less than what we know about ordinary stars. The two collisions of neutrons stars observed in O2 and O3a did not provide a lot of detailed information about the internal structure of neutron stars. More detections will be needed for these investigations.

A surprising consequence of the LIGO/Virgo observations is that the spins (essentially the rotation speeds) of the detected black holes are not large, and their axes not closely aligned.

Most stars rotate, and most binary star systems have stars that rotate along axes parallel to the axis of their common orbit: the spins are aligned. It was thus expected that systems of black holes that were formed together would have locked themselves, aligning their axes of rotation. This lack of alignment suggests that perhaps the black holes were formed separately and they somehow got together to form a binary system.

The spins of the black holes cannot be measured with great precision for each black hole before colliding in the observations made in O1–O3a. Nevertheless, recent results seem to indicate that they are lower than what was expected (based on observations of X-ray binary stars). There is a maximum spin speed, at which the black-hole horizon is rotating at the speed of light. An indicator of black-hole rotation is its spin parameter. Its value can range between 0 (no rotation at all) and 1 (the maximum possible value). X-ray astronomers had observed that black holes with larger masses had higher spins. Before the gravitational wave detection era, it was believed that most black holes with masses larger than 10 solar masses would have spin parameters close to 1. One of the most massive known black holes (Cygnus X-1 with 12 solar masses) has a spin parameter larger than 0.98.

However, the spin parameters observed by LIGO/Virgo have been lower than 0.5, which adds to the mystery of this new family of black holes revealed by gravitational wave detections. A possible explanation for the difference is the fact that black holes detected by their X-ray emissions have accretion disks, like the one depicted in the illustration 5.4, for Cygnus X1. These rapidly rotating disks could have spun up these types of black holes. Theoretical astrophysicists are hard at work to solve these mysteries.

In all cases, the mass of the final object left after the collision is lower than the sum of the masses of the objects that collided. The difference corresponds to energy emitted in the form of gravitational waves. This energy is between 3 percent and 6 percent of the masses in question. In the case of GW190521, the most massive binary black-hole system detected to date, the radiated gravitational energy corresponds to almost 7 solar masses. The quantity of energy emitted per second is equivalent to about 60 times the energy emitted per second by all the stars in the universe in visible light!

Before the gravitational wave discoveries, the existence of binary black-hole systems (and their mergers) was suspected but had not been observed. Now, with so many discoveries, not only have the existence of these systems been proven, but there is also an estimated rate of these events: between 2 and 5 of them per million years per galaxy. That's why we need instruments sensitive to events very far away, with a volume including many millions of galaxies.

By observing gravitational waves, objects with surprising properties are being discovered that were not expected based on previous observations. And this is stimulating new theoretical and astrophysical work on the subject. It is the beginning of a new astronomy and of a new era in astrophysics!

11

The Birth of Gravitational Wave Multi-Messenger Astronomy

The Advanced LIGO detectors gathered data in three observing runs, O1–O3, between 2015 and 2020. During O2, on August 1, 2017, Advanced Virgo joined the LIGO detectors operating in scientific mode, and on August 17, a detection of a collision of two neutron stars was observed. It turned out to be a treasure trove of astrophysical results. We dedicate this chapter to that spectacular event.

11.1 The Network of Detectors

Traditional telescopes can be pointed to any visible direction in the sky to observe a source that is emitting light. Gravitational wave detectors cannot be oriented. They are fixed at the observatories housing them and can receive signals from almost all directions. Therefore, if we detect a wave with a single detector, it is almost impossible to infer where the signal came from.[1] If the signal is received in two detectors exactly at the same time, one can infer it came from some point on an imaginary ring in space located between the detectors. Each point on the ring has the same distance to both detectors, so the signal takes the same time to reach both of them. If the signal arrives at one detector before the other, we can infer that it came from a point on a ring displaced in the direction of the observatory that received it first. For example, the first gravitational wave, GW150914, reached the Livingston Observatory before the Hanford site, and this can give us some idea of where it came from.

A ring in space is still a very poor localization in astronomical terms. In reality, we can do a bit better by using other information in addition to the times of detection (e.g., differences of amplitudes and phases). This tells us that the source is not at any point on the ring but is from a section of it. The diagrams plotting the partial rings on the celestial sphere are colloquially called "bananas" due to their shape; figure 11.1 shows some examples of them. The situation improves qualitatively if we have three detectors. We can then use the three times of arrival to triangulate the position of the source in the sky. This is the same technique used by the police to locate a suspect's cell phone, using the signals received by three or more cell towers. In the case of gravitational waves, the various parameters are known with some uncertainty due to noise. That translates, even with three detectors, to a fairly poor localization in the sky (a blob instead of a banana). More detectors would improve the precision, and fortunately, there will be more detectors joining the network in the near future.

The localization of the source is very important for astronomers. If a precise location in the sky is known, they can point their telescopes in the given direction and look for a signal—optical, gamma rays, radio waves—associated with the source. GW170817, the first collision of neutron stars observed by LIGO and Virgo, was localized within a region that occupied 0.04 percent of the sky. This area is not so small, corresponding to 150 times the area of the sky occupied by a full Moon—but it is the best localized source to date.

Figure 11.1 shows a map with the localization areas of some of the O1 and O2 detections (note their banana-shaped contours). This type of image, in which the celestial sphere is mapped onto

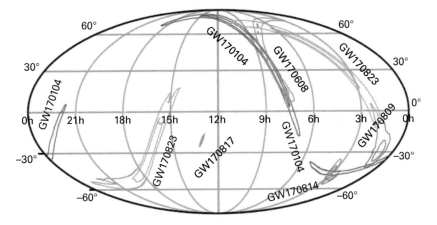

Figure 11.1
Map of the sky showing the areas of higher probability of localization for some sources observed in O1 and O2. (Credit: Adapted from *Physical Review*, X 9, 031040, 2019.)

an ellipse, is common in astronomy, but it distorts the geometry quite a bit. That is why a portion of a ring looks like a banana, and some rings look even more distorted. It is the same problem as found in world maps of the Earth, where the regions near the north and south poles are very distorted. The map of the figure allows to compare the most poorly localized source in O1–O2 (GW170823), obtained with two detectors, and the best localized (GW170817), measured with three detectors.

11.2   Multiple-Messenger Astronomy

The detection of the merger of two neutron stars observed by LIGO and Virgo ushered in a new era in astronomy: multiple-messenger astronomy with gravitational waves. It has been dubbed with such an imposing name because it really involves the use of many couriers, all providing information about the same phenomenon but each contributing to paint a more comprehensive picture than their separate messages alone would deliver.

For example, in 1987, a supernova explosion in the Large Magellanic Cloud (a neighbor galaxy of the Milky Way) was observed with many kinds of telescopes. It also generated neutrinos that were observed in experiments on Earth, giving us rich information about the explosion mechanism. The importance of having multiple messengers is reflected in what happened on August 17, 2017. That day the LIGO and Virgo instruments detected GW170817, a source of gravitational waves emitted by the collision of two small objects that could be neutron stars. It happened at 12:41:04 Coordinated Universal Time, which corresponds to 8:41 a.m. Eastern time. Independently, the gamma ray detectors of the NASA Fermi satellite detected a gamma ray burst (GRB) of short duration (less than 2 sec) that received the name GRB170817A, approximately 1.7 sec after the instant when LIGO and Virgo observed the collision. The European gamma ray telescope INTEGRAL confirmed Fermi's detection in a later reanalysis of its data. Fermi had made its detection public with an automated system, and LIGO/Virgo alerted the astronomical community shortly thereafter. The estimated distance to the source was 130 million light-years, which is close to the Earth by astronomical standards, so electromagnetic waves from the collision could perhaps be observed. This immediately launched an observation campaign with multiple observatories and telescopes, including those from the neutrino and cosmic ray community. The campaign ended up involving more than 2,000 astronomers in addition to the members of the LIGO/Virgo Scientific Collaboration. The telescopes looking for visible light from the blob in the sky had to wait for a few hours until it was nighttime. They concentrated their observations on about 40 galaxies in the region of interest that LIGO/Virgo had found.

Slightly under 11 hours after the alerts went out, the Swope telescope,[2] operated by the 1M2H Collaboration,[3] observed a very bright object that did not appear on previously archived images they had from the same region of the sky (see figure 11.2).

Figure 11.2
Images obtained by the TOROS Collaboration. Left: The kilonova appears next to the galaxy NGC4993. Right: The same source is shown with more digital magnification and the galaxy eliminated for better visualization. Mario Díaz, one of the authors of this book, is the principal investigator of TOROS. (Credit: Adapted from *Astrophysical Journal Letters* 848, L29, 2017.)

The International Astronomical Union assigned this object the identification AT217gfo, located near the galaxy NGC4993, at about 130 million light-years from Earth. This object was independently detected by various groups within the hour (see figure 11.3). Observations continued for several days (e.g., figure 11.2), and for some instruments, months. In the following days, the brightness dimmed in the blue but it increased in redder hues. Nine days later, X-ray emissions were detected and 16 days in the wake of it, radio waves were detected as well. They were all coming from the same direction in the sky. Such emissions lasted longer than a year in the radio band.

The 2017 observation of gravitational waves and a gamma-ray burst produced by the same event provided another confirmation of Einstein's prediction: gravitational waves travel at the speed of light. Since the gamma rays and gravitational waves were detected only 1.7 sec apart from each other after traveling for 130 million years,

Figure 11.3
The various terrestrial and space observatories that attempted to observe GW170817 indicated with light dots on a terrestrial map. The dots outside the map represent the observatories on board satellites also involved in the observation. (Credit: LIGO.)

that provides a very stringent bound on any difference between the speed of gravitational waves and the speed of light (one part in 10^{16}).

11.3   The Mystery of the Gamma Rays

The emission of gamma rays in a collision of neutron stars was not completely unexpected. The origin of GRBs had been a mystery that was still being unraveled, and GW170817 provided important clues.

During October 1963, the US Air Force launched the first of a series of satellites known as Vela ("velar" means "to watch" in Spanish). They received this name since their central mission was to monitor the compliance by the Soviet Union of the 1963 Partial Test Ban Treaty.[4] These satellites had sensors to detect X-rays, gamma rays, and other high-energy particles that could result from a nuclear explosion. During all their years in operation, they never observed a violation of the treaty. However, with the fourth series of these satellites, which had enhanced sensitivity, a very intense source of gamma rays was detected in 1969. But it did not come from the ground: it came from outer space. Aside from the fact that the source was unknown, and the distance to it could not be inferred, one thing was certain: it originated in an extremely energetic event.

This event motivated the design of space missions to start the study of the phenomenon. In 1991, NASA launched the Compton Gamma Ray Observatory, named in honor of the American physicist Arthur Holly Compton, who received the Nobel Prize in 1927. It was the heaviest satellite launched by NASA at the time. Its mission was part of the NASA series of Great Observatories, which also included the famous Hubble Space Telescope. Several other missions launched by NASA and by the European Space Agency were dedicated to studying these events.

The first years of cosmic gamma ray observations allowed a preliminary classification of the various GRBs detected. GRBs could be divided into two categories: long and short duration. In those of long duration, most of the burst energy is emitted over a period longer than 2 sec, and they have a "soft" (i.e., concentrated in longer wavelengths) emission of energy. The short-duration bursts lasted less than 2 sec and emitted most of their energy with a "harder" (shorter wavelength) emission. The origin of the two groups appeared to be different.

In February 1997, it was possible for the first time to identify with certainty the distance to a GRB source: the BeppoSAX satellite—an Italian–Dutch collaboration—detected GRB970228[5] and observed an associated emission of X-rays. Some 20 hours later, the William Herschel telescope in the Canary Islands identified the optical counterpart of the event. It was located in a galaxy so distant that it could barely be observed, and for many years, the distance to it could not be determined precisely. But one thing was clear: these explosions originate outside our galaxy. Later that year, the same satellite identified another source: GRB970508. This was the second GRB observed with an associated emission of X-rays and visible light, and its distance was determined to be 6 billion light-years away.

The following year, the connection of GRB980425 with the explosion of a supernova was established. The first years of the twenty-first century allowed astronomers to determine that the GRBs of long duration and soft energy were associated with the collapse of a massive star, a process in which either a neutron star or a black hole is created. But until 2017, no conclusive evidence identified the progenitors of the short GRBs. These have such high energy levels that in less than 2 sec, they emit all the energy the Sun will produce in its life. Various elements in the observations have led to the belief that the emission of these GRBs is beamed as a narrow jet. It is expected that statistically, only a small fraction of these jets will point toward Earth and be detectable. But it is also estimated that these events are sufficiently infrequent as to happen only twice every million years per galaxy. We should not be so sad about not witnessing such an event in our Milky Way. Their energy is so huge that if a burst like this did occur in our galaxy and the beam were directed toward us, it could most likely extinguish life on Earth. Some scientists have speculated that the massive Ordovician extinction that happened 440 million years ago was provoked by a GRB emission striking straight at our Earth.

The GRB observations during this century raised the suspicion that the short GRBs involved small progenitors compared to the supernovas, like collisions of binary neutron stars, or of a neutron star and a black hole. Immediately before the collision, the stars rupture due to the mutual forces generated.[6] The emission produced by the rupture is very fast because of the small volume of the neutron stars (with a radius of about 10 km) and because in the final moments, matter is moving close to the speed of light. But the evidence was not conclusive. Only with GW170817 was it unambiguously verified that a collision of neutron stars produces gravitational waves and that less than 2 sec later, a gamma ray jet reaches the Earth. Almost immediately, several observations showed a series of effects that had been predicted by various theoretical models proposed by astrophysicist for collisions of neutron stars. The birth of the multi-messenger astronomy had solved a mystery that was more than 50 years old. For both supernova collapses (long GRBs) and the mergers of two neutron stars (short GRBs), the most likely outcome is the creation of a compact object heavier than a neutron star. GRBs are the wailing sounds of black holes being born.

11.4   All That Glitters . . .

The collision of these two neutron stars helped solve another mystery besides that of short GRBs: the origin of the heavy elements that we find in the universe. The heaviest elements that can be formed in stars or novas (or even supernovas) in significant quantities are iron and cobalt. Heavier elements also get produced, but not in quantities large enough to explain, for instance, the abundance of precious metals like gold and platinum found on the Earth.

When the two neutron stars collide, a gigantic explosion ejects matter in an expanding shell of colliding neutrons and atoms moving at 1/3 of the speed of light. This material is compressed at densities that are hard to imagine and completely unachievable on Earth. The blast generates a nuclear process known as the r-process. The r-process entails a succession of rapid neutron captures (hence the "r" in the name) by one or more heavy seed nuclei, such as iron (recall that neutron stars are made of more than neutrons). Most of the elements that are formed through this process of fusion are unstable isotopes that decay radioactively.

It is through this process that electromagnetic waves in the ultraviolet, visible, and infrared regions of the spectrum are emitted. A cocoon of fireworks shines intensely: it is known as a "kilonova," due to the amount of energy being released, which falls between the amount emitted by novas and that emitted by supernovas. Novas are also the product of stellar collisions, but they involve stars that are not so compact (e.g., white dwarfs). There is not as much mass involved as with kilonovas, and the energy released is about 1,000 times lower. Plate 7 shows an artist's conception of a kilonova.

All chemical elements consist of ensembles of neutrons and protons tightly packed in the atomic nucleus with a surrounding cloud of electrons that balances its electric charge (the electric charge of a proton is the same as that of an electron but is of the opposite sign). The simplest of all the elements is a hydrogen atom with only one proton and one electron.

After the Big Bang, the universe expands, cools, and matter starts to form, beginning with elementary particles, followed by atoms and molecules, in a process known as nucleosynthesis. About a few hundred thousand years after the big explosion, the universe is cool enough for the nuclei of atoms formed by neutrons and protons to start capturing electrons and forming atoms. The process starts with hydrogen and a much smaller quantity of the next heavier element, helium. That is why hydrogen and helium are called "primordial" elements. These elements form molecules and group together in big clouds, where stars are created. It is in stars that more complex chemical elements are formed in a fusion process. In particular, all the elements of the periodic table up to cobalt and iron are formed in stars, with others in much smaller quantities. When the stars explode, the created elements eventually make up a newly formed star system like the solar system hosting our Earth and are enough to explain the evolution of life on the planet, including human life. Paraphrasing Carl Sagan, we are made of stardust. Plate 8 shows a periodic table listing the origins of the various elements.

In nova and supernova explosions, the process of creation of heavier elements—like gold and platinum—continues but without producing sufficient quantities of them to explain their abundance ratios in the universe. With the neutron star collisions, the process of creation of all chemical elements is complete. This is what has been learned from the analysis of the observations subsequent to the detection of the collision made by LIGO and Virgo, and it is a good example of what the new discoveries from multi-messenger astronomy can achieve. Although the spectra obtained from AT2017gfo (the kilonova associated with GW170817) were approximately consistent with an outflow of radioactive heavy elements, there was no clear identification of any particular element. But in October 2019, a team of scientists, led by Darach Watson and Camilla J. Hansen from the Niels Bohr Institute in Denmark, reported that after a reanalysis of the spectra obtained in 2017, they managed to identify the presence of the element strontium. This element is known to be the result of a neutron-capture process. It was the smoking gun proving the long-suspected speculation that neutron star mergers, through their kilonova explosions, are where heavy elements are manufactured in the universe. It is estimated that in AT2017gfo, gold and platinum equivalent to 10 times the mass of the Earth were produced and ejected into space during the burst, together with 16,000 Earth masses of other heavy elements. We may be tempted to go and pick up that much gold, but that gold is spread out in the universe. The gold we find on Earth probably comes from similar events.

11.5   The Speed of Expansion of the Universe

Observations using different astronomical instruments of the collision of two neutron stars have contributed to our understanding of the origin of short GRBs and the genesis of heavy elements in the universe. Remarkably, these observations have also allowed astronomers to measure in a completely novel way the rate of expansion of the universe. This is a contentious topic, because the two other existing methods to measure this rate yield results disagreeing with each other.

In section 3.4, we described how in 1929, Hubble discovered that the speed at which all galaxies recede from the Earth is proportional to their distance from Earth. That is, if you divide the numerical value of the velocity of a given galaxy by the numerical value of its distance from the Earth, the result always gives the same number, no matter which galaxy you use for the calculation. The numerical value of the result is known as the Hubble–Lemaître constant. This constant is what astronomers wish to measure.

The velocity of galaxies and other objects can be determined in a relatively straightforward manner through their redshifts. Measuring distances, however, is more complicated. As we discussed in section 2.2, parallax can be used, but only for nearby objects. As the distances increase, the parallax angles decrease and eventually are too small to be measured. For objects farther away, astronomers typically use relations between the objects' brightness and other properties of them, like their spectra or how their intensity varies with time. Hubble used the relationship between the time variation of light in a type of stars known as Cepheids.

The two modern methods were developed by two different collaborations. One is used by the SH0ES collaboration. Its name is an acronym for "Supernova H_0 for the Equation of State" (H_0 is the Hubble–Lemaître constant). It is based on using the relationship between how much light supernovas emit and other properties of the emission, such as how it decays with time (the "light curve"). Once we know how much light they emit, we can infer how far away they are by noting how luminous they appear when observed from the Earth.

The other method is pursued by the Planck collaboration and uses data from the European Space Agency's Planck satellite, named after German Nobel Prize laureate Max Planck, who is one of the creators of quantum theory. This satellite measures very precisely the cosmic microwave background. As already mentioned, this background is composed of electromagnetic waves that have traveled through the expansion of the universe. From the details of this travel, the collaboration can infer the value of H_0.

The collaborations get different results. At the time of writing of this book, the most recent result from SH0ES is 73 km/sec per megaparsec, while Planck obtains a value of 67 km/sec per megaparsec. The observation of systems like GW170817 with electromagnetic and gravitational waves adds another method for measuring H_0 and could eventually help settle the debate. The amplitude and frequency of a gravitational wave contain information about the distance between the source and the Earth. If the system can also be observed optically and its host galaxy identified, we could also infer its velocity (through the redshift of the galaxy). Combining these observations, we could compute H_0, using a method completely independent from those used by SH0ES and Planck. The detection of GW170817 and its localization in the galaxy NGC4993 allowed LIGO/Virgo scientists to do the computation. The value obtained has a large degree of uncertainty. The likely value is between 43 and 115 km/sec per megaparsec. This range includes both the SH0ES and Planck values, so it cannot settle the debate. However, further detections will narrow that range. It is estimated that with 20 observations, one could end up with an accuracy of 2 percent. Multi-messenger astronomy with gravitational waves will help elucidate these great cosmological mysteries.

12

The Future

In addition to the continuous improvement of the LIGO and Virgo detectors, there exists a series of projects and ideas—some under way, others just proposed—to detect gravitational waves from different types of astrophysical sources and with diverse technologies. This chapter summarizes the situation at the time of the writing of this book.

The projects can be divided into four types: additional LIGO-type detectors, longer terrestrial detectors, space-based detectors, and detectors for primordial gravitational waves.

12.1 Growing the Terrestrial Detector Network

As demonstrated by the detection of the 2017 collision of neutron stars, a network of detectors operating together allows more precise localization of the sources. As more detectors are involved, it is more likely that at least three are operating simultaneously for triangulating. A network also lowers the chance of a false positive being coincident in all detectors. This is why the LIGO and Virgo detectors have operated jointly in the past and will continue to do so in the future. Two more detectors will join the network in the near future: KAGRA in Japan and LIGO-India.

KAGRA

The Kamioka Gravitational Wave Detector (KAGRA) is a detector in Japan that is 3 km long and is located in the Kamioka Observatory ("Kamioka" is the name of a mining company) built inside an old tunnel previously used for mining operations. The project is led by the Japanese physicist Takaaki Kajita, the Nobel Prize winner in 2015 for the discovery of neutrino oscillations. At the same site are two other important experiments to detect neutrinos. The advantage of being in an underground tunnel is that it significantly reduces the seismic noise associated with movements of the Earth's surface. The KAGRA instrument also incorporates cryogenic technologies to cool the mirrors of the instrument and therefore reduce their thermal noise. Japan has a long history in the search for gravitational waves, having operated two prototype detectors, TAMA (300 m long) and CLIO, which tested the underground and cryogenic techniques in the Kamioka mine.

Although in its initial stages the KAGRA detector will have less sensitivity than LIGO and Virgo, its addition to the network will improve the sky localization of the detected sources. In the year 2020, KAGRA was supposed to start taking data together with LIGO and Virgo, near the end of O3. However, the Covid pandemic made O3 end a month earlier than anticipated. There was instead a month of joint data taking with the GEO600 and KAGRA detectors in April 2020. The KAGRA detector is expected to take data with LIGO and Virgo in the fourth observing run in 2023.

LIGO India

During the period of Initial LIGO (2002–2010), the Hanford LIGO observatory housed two interferometers: one was 2 km long and the other 4 km long. The plan was that if a signal was detected and both instruments registered it, the shorter interferometer would do so at half the amplitude of the longer one. With the approval of Advanced LIGO, the original project called out for the installation of two 4-km interferometers at Hanford and one at Livingston. Negotiations started simultaneously with Australia and other partners, including India, to construct an observatory and install the second Hanford interferometer there. It would had been the one and only observatory in the southern hemisphere, which would have helped considerably with the localization of signals in the sky. In the end, the Australians abandoned the idea due to financial constraints, and India offered to construct the installations to house it. The second Hanford interferometer was packed in crates instead of installing it, and an agreement was signed between NSF and scientific agencies in India to send it to India and have it installed in an observatory similar to the ones at Hanford and Livingston. LIGO-India suffered some delays. But remarkably, the day after the announcement of the first detection in 2016, Indian Prime Minister Narendra Modi tweeted in support of it, and the process finally began to progress. The final agreement was signed in March 2016. At the time of the writing of this book, the project is still in the planning and organization phase, but the land has been purchased for its installation in the District of Hingoli, in Maharashtra. It is expected that the observatory will start to operate in 2027, although the schedule is evolving. This detector is likely to start to operate as the fifth instrument in the network (along with the two US-based LIGO detectors, Virgo and KAGRA), further optimizing the possible astrophysical discoveries.

12.2 Even More Advanced Detectors

In 2020, LIGO and Virgo had not achieved their design sensitivity, but they had already planned improvements for the future, which will soon be implemented. The slightly upgraded version of Advanced LIGO is called "LIGO A+" (Advanced Plus). It will use more advanced quantum optical techniques to reduce the noise at high and low frequencies. The vacuum squeezing techniques currently used reduce the noise at high frequencies, but they increase the noise at low frequencies (where other noise sources are still larger but are being reduced). The new techniques will deal with this problem. There will also be different mirrors used with lower energy losses in their coatings, which will reduce the thermal noise. This proposal has secured funding at NSF, and the vacuum chambers and civil construction needed for the upgrade are being installed as of this writing. These advances are likely to be used in 2024, if not earlier. Virgo is making similar improvements.

In principle, the current LIGO 4-km facilities could house even more advanced detectors than A+, using technologies that are currently being developed. This would be done using heavier and even less lossy mirrors, cryogenic temperatures, and lasers of different wavelengths. The most mature concept is called LIGO Voyager and is being considered by the community.

The length of the current LIGO/Virgo Observatory facilities present fundamental physical limitations that are not expected to be overcome with improved technologies. To go beyond such limitations, larger detectors are needed: with the same precision to measure distances, an interferometer with 10 km arms is 10 times more sensitive to gravitational waves than one with 1 km arms. Such an improvement implies building new observatories with larger vacuum systems, which requires a considerably larger investment of money. These new observatories are called "third-generation detectors" (Initial LIGO/Virgo were in the first generation, and Advanced LIGO/Virgo in the second). The science that would be done would be qualitatively more advanced. The mergers of black holes could be detected from any time in the history of the universe, providing their cosmic history and proof the of presence (or absence) of primordial black holes. Frequent mergers of neutron stars detected from farther away would be guaranteed to have gamma-ray counterparts, giving us multi-messenger observations about those events. The Hubble–Lemaître constant would be measured with great precision. Many mergers of compact objects would have large enough signal-to-noise ratios to differentiate collisions of small black holes from heavy neutron stars. In collisions of neutron stars, the loud signal would allow measurements of their structure, representing nuclear physics measurements of energies impossible to achieve in Earth laboratories. Loud signals of mergers of black holes would be even more precise tests of general relativity, providing details about their horizon structure and perhaps evidence for quantum gravity theories. The science prospects are so compelling that for many years, two new dectector concepts have been explored and are expected to lead to at least two third-generation detectors, one in Europe and another in the US.

The European project for a third-generation interferometer is called the "Einstein Telescope" and is in advanced planning with the participation of institutions in Belgium, France, Germany, Hungary, Italy, Netherlands, Poland, and the United Kingdom. It is conceived as three V-shaped cryogenic interferometers with arms 10 km long housed in an underground triangular facility. The US concept, "Cosmic Explorer," would be an aboveground L-shaped interferometer 40 km long. Both these concepts would take advantage of heavier mirrors. Artistic drawings of these concepts are shown in figure 12.1.

The need to have a global network of third generation instruments and their high costs (at least a billion dollars for the initial budget of each) will very likely lead to the projects becoming multinational. Since these projects take decades to be designed, built, and operated before yielding results, now is the time to start on the long path to more exciting discoveries.

Figure 12.1
Artist renditions of the Einstein telescope (top panel) and of the Cosmic Explorer
(bottom panel). (Credits: Einstein Telescope, http://www.et-gw.eu; Cosmic Explorer
project, https://cosmicexplorer.org/.)

12.3   Even Longer Detectors: In Space

Astrophysical systems with much larger masses than neutron stars and stellar-mass black holes generate lower frequency gravitational waves with much longer wavelengths that cannot be detected with instruments on Earth.

We know there exist black holes of millions and sometimes billions of solar masses at the center of almost every galaxy. The observations proving the existence of the 4 million solar-mass black hole at the center of the Milky Way was the subject of the 2020 Nobel prize awarded to Andrea Ghez and Richard Genzel, who independently have measured the motion of stars around it for decades. The merger frequency of black holes so large would be in the milli-Hertz to tenths of a Hertz, with wavelengths of millions of kilometers. To measure these, the detectors must be longer. On Earth, issues related to construction, acquisition of land, the curvature of the Earth and seismic noise place severe practical limitations on how long a detector can be. These difficulties lead to the consideration of an alternative: placing the detectors in outer space.

LISA

The Laser Interferometer Space Antenna (LISA) concept is an interferometer in space, sending laser beams between satellites in orbit 2.5 million km apart (the exact length has not yet been chosen). Initially it was a joint project of NASA and the European Space Agency. At present, it is led by the latter, with a smaller contribution from NASA and a launch date of 2034. It consists of three satellites in an equilateral triangle configuration (see figure 12.2). Each satellite has two telescopes to receive the laser light coming from the other two, and two test masses with mirrors that float freely inside each satellite. The triangle of satellites is in an orbit similar to that of the Earth around the Sun and trails the Earth by about 20 degrees, or tens of millions of kilometers from it.

Figure 12.2
Artist's conception of the LISA satellites, in orbit around the Sun following the Earth. A supermassive black hole with its bright accretion disk lurks in the background. (Credit: NASA, public domain.)

On December 3, 2015—the same year of the first LIGO detection—the European Space Agency launched the satellite LISA Pathfinder to test in space the technology needed to keep the masses floating freely in the interior of a satellite. For the masses to float freely, adjustments are needed on the satellite that contains them, due to the impact on it of particles emitted by the Sun (i.e., the solar wind) and other perturbations. This was one of the great technological challenges for the LISA project; this mission demonstrated that the noise due to these adjustments does not compromise the sensitivity of the instrument.

Just like LIGO, LISA has a range of frequencies at which is most sensitive, but it is a much lower one, spanning a band from 1 Hz at the higher end down to about 0.0001 Hz at the lower limit. For this range of frequencies, the expected sources are supermassive black-hole collisions (which would occur with the collision of galaxies), the infall of compact objects in the throes of these behemoths, or waves produced by white dwarfs in our own galaxy. The gravitational waves produced by the collision of supermassive black holes are so intense that LISA could see them essentially from any place in the universe. The loud signals would allow the study of the waves in much more detail than can be achieved with LIGO. Excitingly, a large stellar-mass binary black-hole system could be first detected by LISA when its orbit reaches the frequencies that lie in LISA's bandwidth and then later picked up by LIGO when the objects are in the process of their final approach and collision and the frequencies increase rapidly: this would be the equivalent of seeing an event both in X-rays and at optical wavelengths.

LISA would detect some sources of radiation over periods of months, in contrast to durations of tens of seconds that make up the longest detection times of LIGO. Although LISA is only one instrument, as it detects the waves while orbiting the Sun, the localization in the sky of the source could be disentangled from the waveform. Also, given the distance at which it can observe sources, LISA could study cosmological parameters, such as the rate of expansion of the universe.

Pulsar Timing
The rotation rate of neutron stars, as observed in pulsars (remember the analogy in section 4.7 with the rotating light in a lighthouse), is very stable. In fact, they are comparable with terrestrial clocks in precision: at present, the best terrestrial clocks are about 100 times more accurate than pulsars. The precision of pulsars as clocks is about 100 parts in a billion. If several pulsars are monitored at different locations in the galaxy, a change in the spacing between pulses will indicate the passage of a gravitational wave due to the space-time distortion that the wave produces. This type of

measurement is known as pulsar timing. That is, the Earth–pulsar distance would be used as the giant arm of an interferometer, and the radio waves would play the role that laser light plays in LIGO. A detector with such long arms is sensitive in a range of frequencies even lower than LISA, from a millionth to a billionth of Hertz. In that range of frequencies, we would expect to see evidence of the collisions of supermassive black holes.

Three groups are collaborating on this topic and are participating in the International Pulsar Time Array collaboration. One of the groups is based at the Parkes Observatory in Australia. This observatory is famous for receiving the best television images of the Apollo 11 mission with its radio antenna, popularized in the movie "The Dish." Another group is the European Pulsar Timing Array, which uses several radiotelescopes in the United Kingdom, Germany, Netherlands, Italy, and France. The third group, in the US and Canada, is the collaboration NANOgrav (North American Nanohertz Observatory for Gravitational Waves), which uses the radiotelescopes in Green Bank, West Virginia, and Arecibo in Puerto Rico.[1]

The collaborations have been accumulating data for many years. The main source for this type of instrument is a stochastic background produced by the many collisions of supermassive black holes. The detectability of such a signal increases with the duration of observation, and so in the scientific community, there is the expectation that this technique will detect gravitational waves in the near future.

12.4   Detecting Primordial Gravitational Waves

It is believed that the Big Bang produced primordial gravitational waves. Such waves come in all frequencies and from all directions in the sky. Different theoretical models predict different amplitudes at each frequency. LIGO conducts searches for these types of waves, but up to the writing of this book, it has not detected them, putting an upper bound on the amplitudes in its frequency band between 10 to 1,000 Hz. It is expected that LISA and the pulsar timing experiments will do the same in their respective bands.

Another way of detecting these gravitational waves would be through the imprint they leave on the cosmic microwave background radiation. Several experiments study the background for this purpose. In particular, the BICEP experiment (Background Imaging of Cosmic Extragalactic Polarization) and the related Keck array (named after American oil magnate W. M. Keck, who established a philanthropic foundation under his name) are telescopes at the South Pole that study the cosmic microwave background to look for variations in it, like the ones produced by gravitational waves.

In 2014, the BICEP collaboration announced that it had detected these variations, which generated a lot of excitement throughout the world. But it was soon realized that the result was a mistake. Similar variations in the cosmic microwave background are produced by cosmic dust. This was known, and care had been taken when processing the data to eliminate this effect, but an error in the calculations failed to eliminate it properly. Much progress has been made since then, and even more sensitive instruments are planned, with telescopes operating in a wide range of frequencies to distinguish cosmological signals from dust in the galaxy.

This type of gravitational wave was produced in the very early universe, when it was much smaller than it is now. In the microscopic world, the effects of quantum theory are important. Quantum theory usually applies in the realm of atoms, molecules, and elementary particles and is irrelevant at the astrophysical level where distances are large. But the smallness of the early universe creates a need for quantum physics to explain some aspects of the production of primordial gravitational waves. Therefore, their detection will be the first experimental confirmation of a quantum effect in gravity and would constitute a revolutionary discovery.

Epilogue

Four hundred years ago, Galileo aimed at the sky for the first time an instrument developed to help sailors navigate at sea. He opened a portal to an unknown universe. With it, astronomy started a new era with the capability of observing phenomena never before envisioned. Even today, better and bigger telescopes are being built and planned in the sky and on the ground.

A new type of astronomical instrument, the gravitational wave detector, has now come online, once again opening a new window on the universe. It is a qualitative jump, as when sound was finally incorporated into the silent cinema of the early twentieth century: we can "see" the universe with electromagnetic radiation, and we can "hear" it with gravitational waves. In the short period since the first detection in 2015, we have acquired important new knowledge about our cosmos. But this is only the beginning of gravitational wave astronomy: the scientific community has new challenges, and there will be centuries of different and better instruments, as has happened since the time of Galileo. All this progress has been possible due to the combined efforts of thousands of scientists who have demonstrated a profound spirit of cooperation transcending geographical and cultural boundaries.

We are very proud to have taken part in our own small ways in this new scientific dawn. We are very encouraged to see that new generations of scientists are getting ready to live their own adventures in discovering the mysteries of the cosmos with these new technologies.

Acknowledgments

We thank Reinaldo Gleiser, Dave Reitze, Peter Saulson, and Rainer Weiss for their careful reading of the book and the helpful comments they made on this manuscript. Other colleagues made useful comments about the book, among them Manuela Campanelli, Rodolfo Gambini, Diego Garcia Lambas, Luis Lehner, Carlos Lousto, and Wahltyn Rattray. We thank the many hundreds of colleagues who made the discovery of gravitational waves possible and their work on the future of the field, which is the topic of this book. We have learned from them and enjoyed their camaraderie for many years.

This book grew out of a similar book published in Spanish by the same authors with the collaboration of Lidia Díaz.

This work was supported in part by the US National Science Foundation, the Hearne Institute for Theoretical Physics and the Center for Computation and Technology of Louisiana State University, and the Center for Gravitational Wave Astronomy of the University of Texas Rio Grande Valley. We also thank the Perimeter Institute in Canada and Luis Lehner for their hospitality while we worked on this book.

Notes

Chapter 1

1. Richard S. Westfall, "The Foundations of Newton's Philosophy of Nature." *British Journal for the History of Science*, vol. 1, no. 2 (December 1962), 171–182.

Chapter 2

1. This story is probably apocryphal. It is quoted in Andrew Norton, editor. *Dynamic Fields and Waves*. London: Routledge, 2018.

Chapter 3

1. Albert Einstein. "Über den Einfluß der Schwerkraft auf die Ausbreitung des Lichtes." *Annalen der Physik*, vol. 35 (1911), 898–908. Quote is from page 908.

2. The metric is the distance in space-time.

3. A maximum circle is obtained by slicing the sphere with a plane that goes through its center. On Earth, the equator is a maximum circle, the Tropics (Capricorn or Cancer) are not. Meridians are all maximum circles.

4. Light-year is a measure of distance commonly used in astronomy. It is the distance that a beam of light would travel in a year, approximately 6 trillion miles.

5. The name "Lemaître" was added to this law by the International Astronomical Union in 2018, after having been referred to as "Hubble's law" for many years.

Chapter 4

1. In physics, a moment is the mathematical quantity resulting from the product of a distance (e.g., of a body to some point in space) and a physical quantity associated with this body; so the moment accounts for how the physical quantity is located or arranged in space.

2. We will see in chapter 9 that understanding and characterizing noise for an instrument like a gravitational wave detector is a truly formidable challenge.

3. Roger Penrose. *Shadows of the Mind: A Search for the Missing Science of Consciousness*. Oxford: Oxford University Press, 1996.

Chapter 5

1. A. S. Eddington. *Stars and Atoms*. Oxford: Clarendon Press, 1927, 50.

Chapter 7

1. After Isaacson's retirement from NSF and until Marronetti took over, Beverly Berger was the program officer in charge of Gravitational Physics at NSF, and she was very supportive of the efforts of the Brownsville numerical relativity group.

Chapter 8

1. This situation is difficult to visualize, but imagine skydivers when they form rings in the sky before opening their parachutes.

2. Even today it is not clear what failed in Weber's analysis. For more details, see the book *Gravity's Shadow* by Harry Collins (Chicago: University of Chicago Press, 2004).

3. An interesting coincidence was that in 2016, Fleming Crim, the director of the Mathematical and Physical Sciences Directorate of NSF; and Gabriela González, Dave Reitze, and David Shoemaker, representing LIGO, testified about the discovery of gravitational waves in front of that committee.

Chapter 10

1. Two broad types of operating modes, called "runs," are routinely carried out with the instruments. Engineering runs are mainly to diagnose and correct noises and are denoted E1, E2, and so forth. Science runs are used mainly to attempt to detect gravitational waves. For Initial LIGO, they are denoted S1, S2, and so forth. For Advanced LIGO, they are called "observing runs" and are denoted O1, O2, and so on.

2. UN Resolution A/RES/70/212, dated December 22, 2015.

Chapter 11

1. In reality, the detector has different sensitivities in different directions, so some localization is possible (to about one quarter of the sky), but with great uncertainty.

2. Named in honor of the American astronomer Henrietta Swope and located in Las Campanas, Chile.

3. The name stems from 1M for the diameter of the mirror of the telescopes and 2H for two hemispheres, since the observatory consists of two instruments, one in the northern hemisphere and the other in the southern hemisphere.

4. Its full name was Treaty Banning Nuclear Weapon Tests in the Atmosphere, in Outer Space and Under Water, and it prohibited all test detonations of nuclear weapons except underground tests.

5. This number uses the same date convention that is used for gravitational waves; in fact, it was modeled after the GRB convention.

6. Something similar happens during the collision of a neutron star with a black hole; in that case, only the star ruptures.

Chapter 12

1. Arecibo had been used in the past. Following two breaks in the cables supporting the receiver platform after a storm, the NSF planned to decommission the Arecibo telescope due to safety concerns. On December 1, 2020, the main telescope collapsed before controlled demolition could be conducted. The Very Large Array in New Mexico is being used instead.

Recommended Reading

Marcia Bartusiak. *Einstein's Unfinished Symphony: The Story of a Gamble, Two Black Holes and a New Age in Astronomy*. New Haven, CT: Yale University Press, 2017.

Harry Collins. *Gravity's Shadow*. Chicago: University of Chicago Press, 2004.

Harry Collins. *Gravity's Kiss: The Detection of Gravitational Waves*. Cambridge, MA: MIT Press, 2017.

Harry Collins and Trevor Finch. *The Golem*. Cambridge: Cambridge University Press, 2012.

Jean Eisenstaedt. *The Curious History of Relativity: How Einstein's Theory of Gravity Was Lost and Found Again*. Princeton, NJ: Princeton University Press, 2007.

Fred Jerome. *The Einstein File: The FBI's Secret War against the World's Most Famous Scientist*. Baraka Books, 2018.

Daniel Kennefick. *Traveling at the Speed of Thought: Einstein and the Quest for Gravitational Waves*. Princeton, NJ: Princeton University Press, 2007.

Daniel Kennefick. *No Shadow of a Doubt*. Princeton, NJ: Princeton University Press, 2019.

Abraham Pais. *Subtle Is the Lord*. Reissued in 2005 with a foreword by Roger Penrose. Oxford: Oxford University Press, 2005.

Kip Thorne. *Black Holes and Time Warps: Einstein's Outrageous Legacy*. New York: W. W. Norton, 1995.

Kip Thorne. *The Science of Interstellar*. New York: W. W. Norton, 2014.

Index

Advanced LIGO, 127, 150
AT217gfo, 175

banana plots, 172
bar detectors, 109
BICEP experiment, 195
Big Bang, 50
binary pulsar, 74
black holes, 47, 71, 84
Brownian motion, 144

catalogs of gravitational waves, 163
Chandrasekhar limit, 81
Chapel Hill conference, 70
Copernicus, 7
Cosmic Explorer, 189
Cygnus X-1, 85

dark port, 143
deflection of light, 34
deflection of light by the Sun, 45
Detector Characterization Group, 149

$E=mc^2$, 25
Einstein@home, 93

Einstein equations, 41
Einstein–Rosen paper, 65
Einstein Telescope, 189
end test mass (ETM), 151
expanding universe, 51

Fabry–Pérot cavity, 151
Fermi satellite, 174
fluctuation-dissipation theorem, 146

Galilean physics, 5
general relativity, 29
Grand Challenge, 99
gravitational redshift, 31
gravitational waves, 55, 59
Greek physics, 4
GW150914, 153
GW151226, 161
GW170817, 174

Hanford site, 129
Hertz experiment for gravitational
    waves, 106
Hubble–Lemaître constant, 53, 189
Hulse–Taylor pulsar, 74

Initial LIGO, 120, 151
input test mass (ITM), 151
INTEGRAL, 174
interferometric detectors, 112
interferometry, 123

KAGRA detector, 186
kilonova, 180

Laplace, Pierre-Simon, 13
laser, 124, 137
LIGO, 114
LIGO A+, 188
LIGO India, 186
LIGO Lab, 120
LIGO Scientific Collaboration, 121
LISA, 191
Livingston site, 135
longitude, 13
Lorentz transformation, 24
low tide of general relativity, 68

Maric, Mileva, 28
matched filtering, 91
Maxwell's electromagnetism, 16
Mercury's perihelion precession, 44
metric, 40
Michelson-Morley experiment, 19
multi-messenger astronomy, 171

NANOgrav, 194
National Press Club, 159
neutron stars, 82
Newton's gravity, 1
New York Times, 46, 117, 161
non-Euclidean geometries, 37
numerical relativity, 97

Pathfinder, 192
Physical Review Letters, 158

Planck Colalboration, 182
Pound-Drever-Hall method, 142
power recycling, 143
primordial gravitational waves, 194
principle of equivalence, 30
prototypes (Caltech and MIT), 115
Pulsar Time Array Collaboration, 194
pulsar timing, 193

quadruple pendulum, 146
quadrupole moment, 61

rotating stars, 92

Schwarzschild solution, 48
seismic isolation, 133
SH0ES Collaboration, 182
shot noise, 136
signal recycling, 143
simultaneity, 23
sources of gravitational waves, 89
special relativity, 15
speed of expansion of the universe, 181
speed of light, 17
spin parameters, 178
squeezed vacuum, 138
stellar binary systems, 89
stochastic background of gravitational waves, 94
Superconducting Super Collider, 129
supernova explosions, 93
suspended mirrors, 130
suspension systems, 131
Swope telescope, 174

thermal noise, 144

US Congress, 161

vacuum system, 127
Vela satellites, 176

violin modes, 146
Virgo Project, 121, 152

waves, 10, 58
white dwarfs, 79
Wright–Patterson Air Force Base, 69